1998—2013
上海浦东建筑设计研究院有限公司
作品集

编 委 会

编委会主任：俞　辉　卞能超
编委会副主任：张大伟　管锦饶　强国平　王兆军
编　委　会：孙毓华　张　燕　孙　丰　李林毅　张福绵　张文君　谢　涛
　　　　　　涂秋风　徐　琳　陆　雄　盛棋楸　束顺健　赵国华　徐　建
　　　　　　张　赟　凌宏伟　覃大伟　曹东国　李雪松　林选泉　李永谷
　　　　　　黄晓燕　康秀军　詹华茂　张宝弟
执　行　编　辑：王兆军　徐　琳　祖国庆
图　片　提　供：《浦东设计》杂志编辑部
英　文　翻　译：张永玲
文　字　校　对：付　静
版　式　设　计：上海飞升企业形象策划有限公司

图书在版编目(CIP)数据

上海浦东建筑设计研究院有限公司作品集：1998—2013 / 卞能超,俞辉,王兆军编. -- 南京：东南大学出版社, 2014.1
ISBN 978-7-5641-4634-4

Ⅰ. ①上… Ⅱ. ①卞… ②俞… ③王… Ⅲ. ①建筑设计-作品集-上海市-1998-2013　Ⅳ. ①TU-206

中国版本图书馆CIP数据核字 (2013) 第263057号

上海浦东建筑设计研究院有限公司作品集1998—2013

出版发行	东南大学出版社
社　　址	南京市四牌楼2号　邮编210096
出版人	江建中
网　　址	http://www.seupress.com
电子邮箱	press@seupress.com
经　　销	全国各地新华书店经销
印　　刷	上海豪杰印刷有限公司
开　　本	889 mm×1194 mm　1/12
印　　张	23
字　　数	750千字
版　　次	2014年1月第1版
印　　次	2014年1月第1次印刷
书　　号	ISBN 978-7-5641-4634-4
定　　价	480.00元

本社图书若有印装质量问题，请同营销部联系。电话：025-83791830

前言

当这本沉甸甸的《作品集》编辑完成时,一直惴惴的心绪终于得以释然。细细翻阅时,突然发现许多事情突破了之前的预期,让人自豪而激动。

《作品集》是公司改制重组15年所完成的重要设计成果的一个汇总。从某种意义上说,它就像一股涓涓细流,生动叙说了15年来公司的各种变化和创作细节。这中间,许多项目已经建成,并经受了时间的检验,也有一些项目仍在设计过程之中,一些项目正在进入施工图和施工建造的最后阶段。编辑时,我们限于篇幅,不得不怀着遗珠之憾进行割舍,使最终的《作品集》定稿于建筑设计、市政设计、景观设计、室内设计、规划设计、河道整治6大类131个作品。

这本《作品集》不是单纯简单的作品罗列,而是通过记录的方式,呈现出对多年来设计创作进行的思考,也在讲述着这些项目创作时的一些理念和方式。

对作品的选择,我们特别注重以下三个特征:

一是呈现作品类型的多样性。从公共建筑到商业办公建筑,从文化教育建筑到医疗卫生建筑,从几万平米的住宅到十几万平米的城市综合体,从市政设计到河道整治,从城市设计到城市景观,从系统规划到室内装饰,作品类型的多样性一方面反映出15年来建设市场的繁荣给公司带来的广阔发展空间,另一方面也说明我们的坚持、执着和对市场的适应能力。

二是关注城市特征和地域性。十五年来,公司以"立足浦东、面向上海、走向全国"为经营方针,逐步摆脱了对市内单一市场的依赖,努力构建起全国经营网络。无论是新疆莎车福利中心,还是贵阳中坝路改扩建工程,亦或是西宁湟水森林公园主题区景观工程,都是我们努力争取当地标志性项目的一个缩影。

三是坚持创新,不随波逐流,做自己的建筑,坚持自己的价值取向。不可否认,相对于行业大鳄,我们只属于行业的新生代、实践者,但我们会摒除一些利益,承担起我们应当承担的责任。从浦东市民中心、中环线浦东段,到世博公园,每一个作品都饱含信念,坚守环境意识,追求自身特点,通过灵性更以勤奋留下真实的印记。

我相信,这些作品会随着《作品集》一起写进浦东设计院的历史,意味着曾经的荣耀,也打上了一个年代的烙印。

15年对于一个环境快速变化的年代,是足够长的,足以改变面貌。而这15年,对于设计行业而言,则是最好的时代。基于行业的大背景,我们既迎来的是更广阔的市场,更面对的是复杂的考验。

15年间,公司发生了许多变化——

从初建时年设计产值1000万元逾越至2013年的1.5亿元;

从1家本土的区级设计院到包含多家分支机构的集团性质规模的市级设计机构;

从单一资质到涵盖建筑、道路、桥梁、排水、城市隧道、风景园林、河道整治、工程咨询、城乡规划多领域资质的综合性企业;

从最初的无缘获奖到15年间150多项奖项的累累硕果;

······

这些并不是简单的数字变化,而是市场给予理想和奋斗的回馈。

企业气象缘有人心。这些数字的背后,离不开社会各界的信任支持、鼎力相助、携手陪伴。感谢有你们,在最好的年代,有了精彩的我们。

一个十五年的结束,也意味着新的十五年拉开了序幕。道路绵延,可能是条直线,可能是曲折迂回,也可能是峰回路转。

我们会坚定一路前行,不想说豪言壮语,只请大家拭目以待。

总经理:

2013.12

序言

砺炼品质 熔铸精华

对上海浦东建筑设计研究院有限公司的了解，始于2010年上海世博会项目。当时境内外无数知名设计公司云集上海，竞争激烈，这其中浦东院也表现得非常活跃，在不少项目的投标中都有着不错的表现，出色地完成了上海世博会浦东场地的公共空间设计、上海世博样板组团展馆、世博公园A区(亩中山水)、浦东南路综合改造工程等一大批项目，给我留下了较为深刻的印象。

深入接触之后了解到，上海浦东建筑设计研究院有限公司是在1998年由隶属于上海浦东新区建交委的建筑、城建、园林绿化三家企事业设计单位的基础上改制重组而成，上海浦东建筑设计研究院有限公司这15年的发展历程正处在上海浦东新区开发以及中国改革开放大建设的高潮时期，改制后的浦东院紧紧抓住上海乃至全国城市建设快速发展的机遇，参与设计了许多城市建设及社会公益事业的建设项目，促进了浦东院这些年来突飞猛进的发展。

我认真地翻阅了面前这本"上海浦东建筑设计研究院有限公司15年作品集"样书，作为对浦东院多年来在工程设计行业辛勤耕耘的收获的总结，从中可以看到，他们秉承立足浦东、面向上海、走向全国的发展理念，抱着对设计工作的高度追求，在工业建筑、民用建筑、古建筑、城市规划、市政工程、园林绿化、城市景观、室内装饰设计等的专业领域都取得了可喜的成绩，特别是在住宅、大型公建及配套、特色城镇建设、市政工程、城市(区)景观营造等技术服务中，形成了特有的核心竞争力。而其作品也展示了其公司的理念所示"砺炼品质 熔铸精华"，因此可以说是天时、地利、人和三方面的因素促使浦东院在近十几年时间内发展成为上海乃至全国集建筑设计、市政工程、风景园林、规划设计于一体，具有一定品牌影响力的综合性甲级设计机构。

古语说："不积跬步，无以至千里，不积小流，无以成江海"。这本作品集对公司而言是段小结更是新篇章的起始。未来中国进入新型城镇化的建设时期，市场需求的多元化，绿色建筑理念，设计技术的更新等为工程设计企业的发展提供了新的动力，也对设计企业提出了更高的要求，在此我衷心祝愿上海浦东建筑设计研究院有限公司能继续把握机遇，不断开拓提升，有更多更好的作品奉献给社会！

2013年8月21日
郑时龄　中国科学院　院士
同济大学教授

题词

发扬创新精神
再创事业辉煌

王素卿
二〇一三年十月

王素卿　中国勘察设计协会　理事长

本土创新

崔恺　中国工程院　院士
中国建筑设计大师

二〇二三年九月二十二日

题词

与时俱进

再展宏图

汤志平

二〇一三年十二月十三日

汤志平　上海市城乡建设和交通委员会　主任

因浦东开发而成长
在服务全国中壮大

李佳能

二〇二三年十一月

李佳能 原上海市浦东新区政协 主席

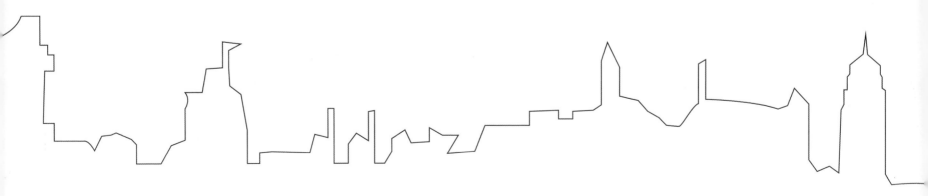

简介

上海浦东建筑设计研究院有限公司(PDAD)于1998年10月在原浦东建筑设计研究院、浦东城建设计公司、浦东园林绿化设计所基础上改制重组而成，现已成为集建筑、市政、风景园林、规划设计四位一体的综合性甲级设计企业。公司现有员工近300名，专业技术人员250多名，其中教授级高级工程师6名，高级工程师70名，各类注册工程师70名。公司拥有一批年富力强、勇于开拓、敢于拼搏的中青年技术骨干，他们经受着海派文化的熏陶，并在众多的海内外设计企业的竞争中锤炼。公司的设计作品创意新颖、贯通中西、精湛厚实、广受欢迎。

目前，公司拥有建筑行业(建筑工程)甲级，市政行业(排水工程、道路工程、桥梁工程、城市隧道工程)专业甲级，风景园林工程设计专项甲级，城乡规划乙级、工程咨询乙级、水利行业(河道整治)等设计资质。公司内设规划方案室、建筑设计一所、建筑设计二所、建筑设计三所、建筑设备所、建筑装饰所、道桥设计所、排水设计所、浦西市政设计所、风景园林设计所、都市空间设计所；拥有建筑、结构、道路、桥梁、风景园林、规划、给排水、电气、暖通、室内装饰、概预算等十多个专业。公司于2000年通过ISO 9001质量体系认证，并连续多年荣获"上海市文明单位"称号。

历年来，公司在工业、民用建筑、古建筑、概念规划、小区规划、市政工程、风景园林、城市景观、室内装饰设计等方面均取得了可喜的成绩，共获得国家级、住建部、上海市等各类设计奖项百余项。

多年来我们一贯秉承"质量第一、精心设计、信誉至上、顾客满意"的企业经营理念，为广大业主提供一流的设计和超值的服务。同时我们也将努力把"PDAD"设计品牌打造成一流工程设计集团！

目录 Contents

建 筑
ARCHITECTURE

办公建筑
Office Buildings

浦东市民中心 (026)
Pudong Civic Center

常州世界贸易中心 (028)
World Trade Center in Changzhou

亚太广场二期 (030)
Second Phase Development Project of Asia Square

包头市人民政府政务服务中心 (032)
Administration and Service Center of Baotou Government

克拉玛依档案中心 (034)
Archive Center of Karamay

南翔财富中心 (036)
Nanxiang Wealth Center

核电秦山二期扩建工程 (038)
Expansion Project of Qinshan Nuclear Power Plant (Second Phase)

金隆大厦 (039)
Jinlong Mansion

商业建筑
Commercial Buildings

花市一条街商业广场 (040)
The Flower Market Avenue Commercial Plaza

太仓永久商业广场 (042)
Taicang Yongjiu Commercial Plaza

宾馆建筑
Hotel Buildings

喀什月星上海城一期（酒店） (044)
The First Phase of Yuexing Shanghai City in Kashi

重庆国港国际酒店 (046)
Guogang International Hotel in Chongqing

文化教育
Cultural & Educational Buildings

谢稚柳陈佩秋艺术馆 (048)
Xie Zhiliu & Chen Peiqiu Art Gallery

上海中学临港分校建设工程 (050)
Construction Project of Lingang Branch
of Shanghai Middle School

上海市川沙中学迁建工程 (052)
Relocation and Construction Project
of Shanghai Chuansha Middle School

上海市临港科技学校 (054)
Lingang School of Science and Technology in Shanghai

医疗卫生
Health Care Buildings

西宁市残疾人康复中心 (056)
Xining Rehabilitation Center for Disabled People

上海市航头基地社区卫生服务中心工程 (058)
Community Health Service Center of Hangtou Region in Shanghai

莎车县疾控中心 (060)
Shache Disease Control Center

莎车福利中心 (062)
Shache Welfare Center

川沙新镇福利院 (064)
New Town Welfare House of Chuansha

交通建筑
Transportation Buildings

诸暨铁路新客站 (066)
Zhuji New Railway Station

产业园及工业建筑
Industrial Park & Industrial Buildings

张家港沙洲湖科技创新园建设项目 (068)
Construction Project of Shazhouhu Science
and Technology Innovation Park in Zhangjiagang

上海榕榕文化科技园概念规划及建筑方案设计 (070)
Conceptual Planning and Architectural Design of
Shanghai Rongrong Culture, Science and Technology Park

江苏省海门市临江新区国际中小企业科技园 (072)
Linjiang New District International Small and
Medium-Sized Enterprise Science
and Technology Park in Haimen, Jiangsu Province

杰事杰合肥产业化基地 (074)
Industrial Base of Jieshijie in Hefei

住宅
Residential Buildings

上海市保障性住房三林基地 (076)
Affordable Housing of Shanghai in Sanlin Region

绿庭尚城 (078)
Lvting Residential District

爱家金河湾住宅区 (080)
Aijia Jinhewan Residential District

上海浦东胡巷村 (082)
Huxiang Village of Pudong in Shanghai

综合整治
Comprehensive Renovation Projects

解放路沿线建筑、景观、道路改造工程 (084)
Renovation Project of Buildings, Roads and Landscape along Jiefang Road

重庆松青路干道环境综合改造工程 (086)
Comprehensive Environmental Renovation Project of Songqing Road in Chongqing

衢州市上下街道改造工程 (088)
Construction Project of Shangxia Street in Quzhou

其他
Other Projects

浦东沪新村教堂 (090)
Huxin Village Church in Pudong District

鹤鸣楼 (092)
Heming Tower

市 政
MUNICIPAL

快速路
Expressway Projects

上海市城市外环线环南一大道工程 (096)
Huannan No.1 Road of the Outer Ring in Shanghai

迎宾大道工程 (098)
Usher Avenue Project

浦东新区罗山路延长线(外环线—龙东大道) (100)
Extended Part of Luoshan Road (from the Outer Ring to Longdong Avenue) in Pudong New District

五洲大道(翔殷路越江隧道—外环线)新建工程 (102)
New Construction Project of Wuzhou Avenue (from Xiangyin Road Crossriver Tunnel to the Outer Ring)

中环线浦东段(上中路越江隧道—申江路)新建工程　(104)
New Construction of the Middle Ring in Pudong New District
(from Shangzhong Road Crossriver Tunnel to Shenjiang Road)

贵阳市中坝路改扩建工程　(106)
Renovation and Extension Project of Zhongba Road in Guiyang

主干路
Main Trunk Road Projects

金海路(日新路—川南奉公路)新建工程　(108)
New Construction of Jinhai Road (Rixin Road to Chuan Nanfeng Highway)

周邓公路(申江路—南六公路)新建工程　(110)
New Construction Project of Zhoudeng Road
(from Shenjiang Road to Nanliu Road)

杨高路(成山路—外高桥5号门)改建工程　(112)
Renovation Project of Yanggao Road
(from Chengshan Road to Waigaoqiao Gate No. 5)

山东省东营市西四路、济南路综合改造工程　(114)
Comprehensive Renovation Project of Xisi Road
and Jinan Road in Dongying, Shandong Province

浙江省德清县北湖街延伸工程　(116)
Extension Construction Project of Beihu Avenue in Deqing County, Zhejiang Province

江苏省通州市"两路一桥"工程　(118)
Double-roads-and-one-bridge Project of Tongzhou, Jiangsu Province

青浦大道(318国道—五浦路)新建工程　(120)
New Construction Project of Qingpu Avenue (from the National Highway 318 to Wupu Road)

浦建路(杨高南路—龙阳路)、沪南路(前程路—区界)改建工程　(122)
Reconstruction Project of Pujian Road (from South Yanggao Road to
Longyang Road) and Hunan Road (from Qiancheng Road to the District Border)

次干路和支路
Trunk Road And Branch Projects

东方路改扩建工程　(124)
Reconstruction and Expansion Construction Project of Dongfang Road

莎车县城南新区古勒巴格路、站前路(南环路)道路工程　(126)
Construction Project of Gulebage Road and
Zhanqian Road (the South Ring) in Nanxin District of Shache Town

合肥工业园区道路　(128)
Road Construction Project of Hefei Industrial Park

外高桥新市镇配套工程　(130)
Infrastructure Construction Project of the New Town in Waigaoqiao District

长清路(成山路—杨思路)扩建工程　(132)
Expansion Construction Project of Changqing Road (from Chengshan Road to Yangsi Road)

桥梁
Bridge Projects

苏州河武宁路桥景观改造工程　(134)
Landscape Renovation Project of Wuning Road Bridge over Suzhou River

苏州河宝成桥景观改造工程　(136)
Landscape Renovation Project of Baocheng Bridge over Suzhou River

浦东软件园陆家嘴分园步行桥工程 (138)
Construction Project of the Pedestrain Bridges
in Lujiazui Subzone of Pudong Software Industrial Park

山东省东营市东三路生态廊道桥梁工程 (140)
Construction Project of the Bridges in the Ecological Corridor along
Dongsan Road in Dongying, Shandong Province

吕梁市离石区东城新区东川河桥梁工程 (142)
Construction Project of the Bridge over Dongchuan River
in the Eastern New Zone of Lishi District of Lvliang

排水
Drainage Projects

曹路镇第二期自然村落污水纳管工程 (144)
The Second Phase Sewage-into-sewer Project of the Existing Villages in Caolu Town

风景园林
LANDSCAPE ARCHITECTURE

公共空间景观
Public Space Landscape

世博公园A块(亩中山水)景观工程 (148)
Landscape Project of Zone A of the Expo Park:
Mountains and Rivers in Zone A

上海滨江森林公园二期工程 (150)
The Second Phase Development Project of Shanghai
Waterfront Forest Park

江苏省徐州市行政中心市民中心广场绿化环境景观工程 (152)
Virescence and Landscape Project of the Plaza of Administration
and Civic Center in Xuzhou, Jiangsu Province

海门市民公园规划设计 (154)
Urban Design of Haimen Civic Park

山东东营西湖景观工程 (156)
Landscape Construction of the West Lake of Dongying in Shandong

梅园公园改造工程 (158)
Renovation Project of Meiyuan Park

长风工业园地块公共绿地建设项目 (160)
Construction Project of the Public Green Spaces in Changfeng Industrial Park

张衡公园 (162)
Zhangheng Park

滨河景观
Riverfront Landscape

江西鹰潭市滨江公园二期景观工程　（164）
The Second Phase Landscape Project of
Yingtan Waterfront Park in Jiangxi Province

海门市圩角河两侧绿化景观建设项目　（166）
Landscape Design of the Riverfront of Weijiao River in Haimen

外滩滨江绿地景观工程　（168）
Landscape Project of the Waterfront Green Spaces of the Bund

东营市东三路河生态廊道（德州路—东营河）景观工程　（170）
Landscape Construction Project
of the Ecological Corridor Along Dongshanlu River in
Dongying (from Dezhou Road to Dongying River)

居住区及商业办公景观
Landscape Project of Residential District & Commercial Office

三林世博家园公共绿地建设工程　（172）
Construction Project of the Public Green Space in Sanlin Expo Home

保利湖畔阳光苑景观工程　（174）
Landscape Project of the Lakefront Sunlight Garden
Developed by Baoli Real Estate

南通东郊庄园　（176）
Eastern Suburb Manor in Nantong

湖南电力科技园景观工程　（178）
Landscape Project of the Electrical Science
and Technology Park in Hunan

永达大厦景观绿化工程　（180）
Landscape and Virescence Project of Yongda Building

大渡口H21—1经济适用房项目绿化景观工程设计　（182）
Virescence and Landscape Project of Affordable Housing
in Zone H21-1 of Dadukou District

江苏省泰州市济川医药工业园(一期)景观工程　（184）
The First Phase Landscape Project of Jichuan
Medicine Industry Park in Taizhou, Jiangsu Province

道路景观
Road Landscape

浦东南路道路景观工程　（186）
Landscape Project of South Pudong Road

中环线浦东段新建及两侧绿化工程　（188）
New Construction and Virescence Project of
the Middel Ring in Pudong New District

临港新城申港大道景观工程　（190）
Landscape Project of Shengang Avenue in Lingang New Town

东方路景观　（192）
Landscape of Dongfang Road

江苏海门北京路景观工程　（194）
Landscape Project of Beijing Road in Haimen, Jiangsu Province

五洲大道(浦东北路—外环线)道路绿化工程　（196）
Road Virescence Project of Wuzhou Avenue
(from North Pudong Road to the Outer Ring)

室内装饰
INTERIOR

商场类
Mall

上海五角场万达商业广场地下一层购物中心室内装饰工程　(200)
Interior Decoration Project of the Shopping Center on
the Ground Floor of Wanda Plaza in Wujiaochang District, Shanghai

上海周浦万达广场商业步行街室内装修设计　(202)
Interior Decoration Design of Commercial
Walking Street of Wanda Plaza in Zhoupu District, Shanghai

上海宝山万达广场室内商业步行街公共空间室内设计　(204)
Interior Public Space Design of Commercial Walking Street of
Wanda Plaza in Baoshan District, Shanghai

南京建邺万达商业广场商业步行街室内设计　(206)
Interior Design of the Commercial Walking
Street of Jianye Wanda Plaza in Nanjing

绍兴柯桥万达广场室内商业步行街室内设计　(208)
Interior Design of the Commercial Walking Street of Keqiao
Wanda Plaza in Shaoxing

长春宽城万达购物中心步行街公共空间室内设计　(210)
Interior Public Space Design of the Walking Street
of Wanda Shopping Center in Kuancheng, Changchun

南通中央商务区A-04地块项目室内设计　(212)
Interior Design of A-04 Zone in Central Business District of Nantong

盐城中南·世纪城(2A地块)商场公共部位室内设计　(214)
Interior Public Space Design of
Zhongnan - Century City Market in Yancheng (2A Zone)

长沙奥克斯广场步行街公共空间室内设计　(216)
Interior Public Space Design of Walking Street of AUX Plaza In Changsha

上海嘉定方舟广场步行街公共空间设计　(218)
Interior Public Space Design of Walking Street
of Fangzhou Plaza In Jiading, Shanghai

南通五洲国际广场公共空间装修设计　(220)
Interior Public Space Decoration Design
of Wuzhou International Plaza of Nantong

酒店餐饮类
Hotel & Food and Drink Shop

上海爱法小天地酒店式公寓(吉瑞商务酒店)室内设计　(222)
Interior Decoration Design of "Shanghai Aifa: Little World"
Service Apartment (J-real Residence Suites)

新疆宏泰房产开发旅游宾馆室内装修设计　(224)
Interior Decoration Design of the Tourism Hotel Developed by
Hongtai Real Estate in Xinjiang

办公类
Office Buildings

上海张江泰豪智能电气有限公司办公楼室内设计　(226)
Office Building Interior Decoration Design
of Shanghai Zhangjiang Tellhow Intelligent Engineering Co., Ltd.

上海奉贤绿庭国际中心写字楼室内装修设计　(228)
Interior Decoration Design of Office Buildings of Lvting
International Center in Fengxian District, Shanghai

住宅类及售楼处
Residential & Sales Office

上海爱法奥朗别墅二期样板房设计　(230)
Model House Design of the Second Phase
Development of Aifa: Orang Villa in Shanghai

合肥吉瑞泰盛国际生活广场售楼处及样板房室内设计　(232)
Sales Office and Model House Interior Decoration Design
of Jerry Tightsen International Living Mall in Hefei

上海松江绿庭广场高层公寓、会所、公寓式办公楼精装修设计　(234)
Refined Decoration Design of Apartments, Clubs and Service Apartments
of Lvting Plaza in Songjiang District, Shanghai

上海绿庭尚城住宅精装修户型室内设计　(236)
Interior Refined Decoration Design of Shanghai Lvting Residential District

城市规划
URBAN PLANNING

中国2010年上海世博会园区浦东场地公共空间规划设计　(240)
Public Space Planning of Pudong Zone for Shanghai Expo Region, 2010

山西吕梁北川河片区修建性详细规划　(242)
Site Detailed Planning of Hepian Area of Beichuan, Lvliang, Shanxi

洪先路金融文化街概念性规划设计及建筑方案设计　(243)
Conceptual Planning and Architectural Design of Financial
and Cultural Streetscape along Hongxian Road

浙江省诸暨市行政中心概念规划及城市设计　(244)
Conceptual Planning and Urban Design of Zhuji Civic
Center in Zhejiang Province

西宁市湟中县职业教育多巴新校区修建性详细规划　(245)
Site Detailed Planning of Duoba New Campus for
Vocational Education in Huangzhong County of Xining

西宁市城市绿地系统规划　(246)
Urban Green Space System Planning of Xining

浦东新区道路建设"十二五"规划　(247)
The Twelfth Five-year Plan of Road
Construction in Pudong New District

浦东新区航头拓展大型居住社区雨污水系统专业规划　(248)
Rainwater and Sewage System Planning of Hangtou Expanded
Major Community in Pudong New District

上海唐镇新市镇核心区公共空间规划设计导则　(249)
Design Guidelines of Public Space Planning in the
New Central Region of Tang Town, Shanghai

西宁市城北区小桥东片区控制性详细规划　(250)
Regulatory Detailed Plan of
Xiaoqiao East Zone in North City District of Xining

西宁市北川河景观风貌概念规划　(251)
Conceptual Planning of Landscape
and Scenery Along Beichuan River in Xining

河道整治
WATERWAY DREDGING

外高桥南块微电子产业基地西区二期水系建设工程　(254)
Second Phase Water Course Construction Project of West Zone
of Waigaoqiao Southern Microelectronic Industrial Base

浦东北路河(洲海路—椿树浦)河道新建工程　(255)
New Construction Project of North Pudong Road Channel
(from Zhouhai Road to Chunshupu River)

椿树浦(浦东北路—张杨北路)河道整治工程　(256)
Comprehensive Renovation Project of Chunshupu River (from North
Pudong Road to North Zhangyang Road)

春塘河(川杨河—世博家园)河道综合整治工程　(257)
Comprehensive Renovation Project of Chuntang
River (from Chuanyang River to Expo Home)

中汾泾(川杨河—杨高南路)河道综合整治工程　(258)
Comprehensive Renovation Project of Zhongfenjing River
(from Chuanyang River to South Yanggao Road)

中心河(川杨河—华夏西路)河道综合整治工程　(259)
Comprehensive Renovation Project of the
Central River (from Chuanyang River to West Huaxia Road)

大寨河(高青路—华夏西路)河道综合整治工程　(260)
Comprehensive Renovation Project of Dazhai River (from
Gaoqing Road to West Huaxia Road)

大寨河(川杨河—高青路)河道综合整治工程　(261)
Comprehensive Renovation Project of Dazhai River (from Chuanyang River to Gaoqing Road)

薛家浜(大寨河—沪南公路)河道综合整治工程　(262)
Comprehensive Renovation Project of Xuejiabang River (from Dazhai River to Hunan Road)

中心河(白莲泾—川杨河)河道综合整治工程　(263)
Comprehensive Renovation Project of the Central River (from Bailianjing River to Chuanyang River)

工程咨询
ENGINEERING CONSULTING

工程咨询项目　(266)
Engineering Consulting Project

附录
Appendix

上海浦东建筑设计研究院有限公司部分获奖项目
Part of Prize-winning Projects Made by PDAD

建筑　　**ARCHITECTURE**

026 建筑
ARCHITECTURE

建筑 ARCHITECTURE

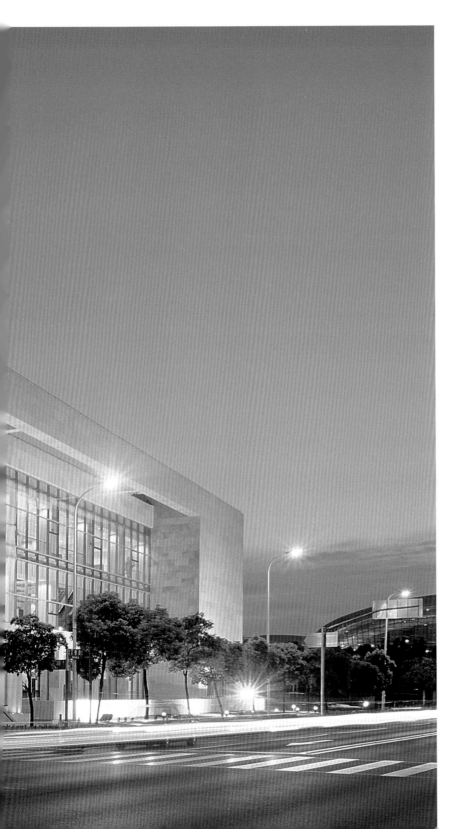

浦东市民中心
Pudong Civic Center

项目地点: 上海市浦东新区　**项目设计时间:** 2005年
项目规模: 17400平方米　**重要奖项:** 上海市优秀勘察设计一等奖
Location: Pudong New District, Shanghai Municipality　**Dates:** 2005　**Scale:** 17400 sq.m.
Major award: First-class Prize of Shanghai for Outstanding Exploration and Design

029 建筑 ARCHITECTURE

常州世界贸易中心
World Trade Center in Changzhou

项目地点: 江苏省常州市 **项目设计时间:** 2007年 **项目规模:** 139600平方米
Location: Changzhou, Jiangsu Province **Dates:** 2007 **Scale:** 139600 sq.m.

030 | 建筑 ARCHITECTURE

亚太广场二期
Second Phase Development Project of Asia Square

项目地点：江苏省昆山市　**项目设计时间：**2007年　**项目规模：**84500万平方米
Location: Kunshan, Jiangsu Province　**Dates:** 2007　**Scale:** 84500 sq.m.

032 建筑 ARCHITECTURE

包头市人民政府政务服务中心
Administration and Service Center of Baotou Government

项目地点: 内蒙古自治区包头市　**项目设计时间:** 2010年　**项目规模:** 47000平方米
Location: Baotou, Inner Mongolia Autonomous Region　**Dates:** 2010　**Scale:** 47000 sq.m.

034 建筑 ARCHITECTURE

建筑 ARCHITECTURE

克拉玛依档案中心
Archive Center of Karamay

项目地点：新疆维吾尔自治区克拉玛依市　**项目设计时间：**2011年　**项目规模：**30000平方米
Location: Karamay, Xinjiang Uygur Autonomous Region　**Dates:** 2011　**Scale:** 30000 sq.m.

036 建筑 ARCHITECTURE

037 | 建筑 ARCHITECTURE

南翔财富中心
Nanxiang Wealth Center

项目地点：上海市嘉定区　**项目设计时间：**2011年　**项目规模：**120000平方米
Location: Jiading District, Shanghai Municipality　**Dates:** 2011　**Scale:** 120000 sq.m.

核电秦山二期扩建工程
Expansion Project of Qinshan Nuclear Power Plant (Second Phase)

项目地点: 浙江省海盐市　**项目设计时间:** 2005年　**项目规模:** 63500平方米
Location: Haiyan, Zhejiang Province　**Dates:** 2005　**Scale:** 63500 sq.m.

金隆大厦
Jinlong Mansion

项目地点: 上海市浦东新区 **项目设计时间:** 2004年
项目规模: 36000平方米
Location: Pudong New District, Shanghai Municipality **Dates:** 2004
Scale: 36000 sq.m.

041 建筑 ARCHITECTURE

花市一条街商业广场
The Flower Market Avenue Commercial Plaza

项目地点: 上海市奉贤区　**项目设计时间:** 2004年　**项目规模:** 20000平方米
Location: Fengxian District, Shanghai Municipality　**Dates:** 2004　**Scale:** 20000 sq.m.

太仓永久商业广场
Taicang Yongjiu Commercial Plaza

项目地点: 江苏省太仓市 **项目设计时间:** 2007年 **项目规模:** 40000平方米
Location: Taicang, Jiangsu Province **Dates:** 2007 **Scale:** 40000 sq.m.

044 建筑 ARCHITECTURE

045 建筑 ARCHITECTURE

喀什月星上海城一期（酒店）
The First Phase of Yuexing Shanghai City in Kashi

项目地点：新疆维吾尔自治区喀什市　**项目设计时间：**2010年　**项目规模：**95000平方米
Location: Kashi, Xinjiang Uygur Autonomous Region　**Dates:** 2010　**Scale:** 95000 sq.m.

046 建筑 ARCHITECTURE

重庆国港国际酒店
Guogang International Hotel in Chongqing

项目地点：重庆市大渡口区　**项目设计时间：**2010年　**项目规模：**39700平方米
Location: Dadukou District, Chongqing Municipality　**Dates:** 2010　**Scale:** 39700 sq.m.

048 建筑 ARCHITECTURE

049 建筑 ARCHITECTURE

谢稚柳陈佩秋艺术馆
Xie Zhiliu & Chen Peiqiu Art Gallery

项目地点：上海市浦东新区临港新城　**项目设计时间：**2011年
项目规模：15000平方米　**重要奖项：**2012上海市优秀工程咨询成果三等奖
Location: Lingang New Town, Pudong New District, Shanghai Municipality　**Dates:** 2011　**Scale:** 15000 sq.m.
Major award: Third-class Prize of Shanghai for Outstanding Construction Project Consulting in 2012

050 建筑 ARCHITECTURE

上海中学临港分校建设工程
Construction Project of Lingang Branch of Shanghai Middle School

项目地点: 上海市浦东新区 **项目设计时间:** 2007年 **项目规模:** 73200平方米
Location: Pudong New District, Shanghai Municipality **Dates:** 2007 **Scale:** 73200 sq.m.

052 建筑 ARCHITECTURE

建筑 ARCHITECTURE

上海市川沙中学迁建工程
Relocation and Construction Project of Shanghai Chuansha Middle School

项目地点：上海市浦东新区　**项目设计时间：**2011年　**项目规模：**81000平方米
Location: Pudong New District, Shanghai Municipality　**Dates:** 2011　**Scale:** 81000 sq.m.

上海市临港科技学校
Lingang School of Science and Technology in Shanghai

项目地点: 上海市浦东新区 **项目设计时间:** 2009年 **项目规模:** 66100平方米
Location: Pudong New District, Shanghai Municipality **Dates:** 2009 **Scale:** 66100 sq.m.

057 建筑 ARCHITECTURE

西宁市残疾人康复中心
Xining Rehabilitation Center for Disabled People

项目地点: 青海省西宁市 **项目设计时间:** 2012年 **项目规模:** 19000平方米
重要奖项: 第五届上海市建筑学会创作奖
Location: Xining, Qinghai Province **Dates:** 2012 **Scale:** 19000 sq.m.
Major award: The Fifth Creation Award of the Architectural Society of Shanghai

058 建筑
ARCHITECTURE

059 建筑 ARCHITECTURE

上海市航头基地社区卫生服务中心工程
Community Health Service Center of Hangtou Region in Shanghai

项目地点：上海市浦东新区航头镇　**项目设计时间：**2011年　**项目规模：**14000平方米
Location: Hangtou Town, Pudong New District, Shanghai Municipality　**Dates:** 2011　**Scale:** 14000 sq.m.

060 建筑 ARCHITECTURE

莎车县疾控中心
Shache Disease Control Center

项目地点：新疆维吾尔自治区喀什地区莎车县
项目设计时间：2013年
项目规模：6500平方米
Location: Shache, Kashi, Xinjiang Uygur Autonomous Region
Dates: 2013
Scale: 6500 sq.m.

建筑 ARCHITECTURE

062

莎车福利中心
Shache Welfare Center

项目地点： 新疆维吾尔自治区喀什地区莎车县　**项目设计时间：** 2011年
项目规模： 19000平方米
重要奖项： 上海市优秀工程设计三等奖
Location: Shache, Kashi, Xinjiang Uygur Autonomous Region　**Dates:** 2011
Scale: 19000 sq.m.
Major award: Third-class Prize of Shanghai for Outstanding Construction and Design

063 建筑 ARCHITECTURE

064 建筑 ARCHITECTURE

065 建筑 ARCHITECTURE

川沙新镇福利院
Welfare House of Chuansha New Town

项目地点: 上海市浦东新区 **项目设计时间:** 2006年 **项目规模:** 11000平方米
重要奖项: 2009上海市优秀工程设计三等奖

Location: Pudong New District, Shanghai Municipality **Dates:** 2006 **Scale:** 11000 sq.m.
Major award: Third-class Prize of Shanghai for Outstanding Construction and Design in 2009

067 建筑 ARCHITECTURE

诸暨铁路新客站
Zhuji New Railway Station

项目地点： 浙江省诸暨市　**项目设计时间：** 2005年　**项目规模：** 17500平方米
重要奖项： 上海市优秀勘察设计三等奖
Location: Zhuji, Zhejiang Province　**Dates:** 2005　**Scale:** 17500 sq.m.
Major award: Third-class Prize of Shanghai for Outstanding Exploration and Design

建筑 ARCHITECTURE

张家港沙洲湖科技创新园建设项目
Construction Project of Shazhouhu Science and Technology Innovation Park in Zhangjiagang

项目地点: 江苏省张家港市 **项目设计时间:** 2012年 **项目规模:** 250000平方米
Location: Zhangjiagang, Jiangsu Province **Dates:** 2012 **Scale:** 250000 sq.m.

069 建筑 ARCHITECTURE

上海榕榕文化科技园概念规划及建筑方案设计
Conceptual Planning and Architectural Design of
Shanghai Rongrong Culture, Science and Technology Park

项目地点：上海市闵行区　**项目设计时间：**2012年　**项目规模：**200000平方米
Location: Minhang District, Shanghai Municipality　**Dates:** 2012　**Scale:** 200000 sq.m.

建筑 ARCHITECTURE

江苏省海门市临江新区国际中小企业科技园
Linjiang New District International Small and Medium-Sized Enterprise Science and Technology Park in Haimen, Jiangsu Province

项目地点： 江苏省海门市　**项目设计时间：** 2012年　**项目规模：** 550000平方米
Location: Haimen, Jiangsu Province　**Dates:** 2012　**Scale:** 550000 sq.m.

073 建筑 ARCHITECTURE

建筑 ARCHITECTURE

 075 建筑 ARCHITECTURE

杰事杰合肥产业化基地
Industrial Base of Jieshijie in Hefei

项目地点：安徽省合肥市开发区　**项目设计时间：**2006年　**项目规模：**87200平方米
Location: Development zone of Hefei, Anhui Province　**Dates:** 2006　**Scale:** 87200 sq.m.

076 | 建筑 ARCHITECTURE

上海市保障性住房三林基地
Affordable Housing of Shanghai in Sanlin Region

项目地点：上海市浦东新区　**项目设计时间：**2008年　**项目规模：**230000平方米
重要奖项：上海市"我最喜欢的房型"奖
Location: Pudong New District, Shanghai Municipality　**Dates:** 2008　**Scale:** 230000 sq.m.
Major award: Prize of Shanghai for "The Most Popular Room Type Design"

建筑 ARCHITECTURE

绿庭尚城
Lvting Residential District

项目地点: 上海市松江区 **项目设计时间:** 2007年 **项目规模:** 400000平方米（一期、二期）
Location: Songjiang District, Shanghai Municipality **Dates:** 2007 **Scale:** 400000 sq.m. (first phase plus second phase)

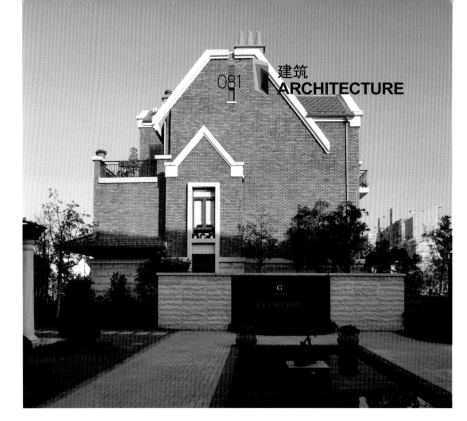

081 建筑 ARCHITECTURE

爱家金河湾住宅区
Aijia Jinhewan Residential District

项目地点：江苏省无锡市　**项目设计时间：**2007年
项目规模：290000平方米　**重要奖项：**2010上海市优秀住宅工程小区设计二等奖
Location: Wuxi, Jiangsu Province　**Dates:** 2007　**Scale:** 290000 sq.m.
Major award: Second-class Prize of Shanghai for Outstanding Residential District Planning and Design in 2010

083 建筑 ARCHITECTURE

上海浦东胡巷村
Huxiang Village of Pudong in Shanghai

项目地点: 上海市浦东新区 **项目设计时间:** 2009年 **项目规模:** 32000平方米
Location: Pudong New District, Shanghai Municipality **Dates:** 2009 **Scale:** 32000 sq.m.

解放路沿线建筑、景观、道路改造工程
Renovation Project of Buildings, Roads and Landscape along Jiefang Road

项目地点: 江苏省海门市　**项目设计时间:** 2007年　**项目规模:** 3000平方米
Location: Haimen, Jiangsu Province　**Dates:** 2007　**Scale:** 3000 sq.m.

085 建筑 ARCHITECTURE

建筑
ARCHITECTURE

重庆松青路干道环境综合改造工程
Comprehensive Environmental Renovation Project of Songqing Road in Chongqing

项目地点: 重庆市大渡口区　**项目设计时间:** 2010年　**项目规模:** 550000平方米
Location: Dadukou District, Chongqing Municipality　**Dates:** 2010　**Scale:** 550000 sq.m.

建筑 ARCHITECTURE

衢州市上下街道改造工程
Construction Project of Shangxia Street in Quzhou

项目地点： 浙江省衢州市　**项目设计时间：** 2012年　**项目规模：** 2.5千米
Location: Quzhou, Zhejiang Province　**Dates:** 2012　**Scale:** 2.5 km

089 建筑 ARCHITECTURE

090 建筑 ARCHITECTURE

浦东沪新村教堂
Huxin Village Church in Pudong District

项目地点：上海市浦东新区　**项目设计时间：**2003年　**项目规模：**873平方米
Location: Pudong New District, Shanghai Municipality　**Dates:** 2003　**Scale:** 873 sq.m.

091 建筑 ARCHITECTURE

093 建筑 ARCHITECTURE

鹤鸣楼
Heming Tower

项目地点： 上海市浦东新区
项目设计时间： 1992年
项目规模： 4200平方米
Location: Pudong New District, Shanghai Municipality
Dates: 1992
Scale: 4200 sq.m.

市政　　MUNICIPAL

096 市政
MUNICIPAL

097 | 市政 MUNICIPAL

上海市城市外环线环南一大道工程
Huannan No.1 Road of the Outer Ring in Shanghai

项目地点： 上海市浦东新区　**项目设计时间：** 1997年　**项目规模：** 25.3千米×100米　**重要奖项：** 2003上海市优秀勘察设计一等奖
Location: Pudong New District, Shanghai Municipality　**Dates:** 1997　**Scale:** 25.3km × 100m
Major award: First-class Prize of Shanghai for Outstanding Exploration and Design in 2003

098 市政
MUNICIPAL

迎宾大道工程
Usher Avenue Project

项目地点：上海市浦东新区 **项目设计时间**：1997年
项目规模：4.4千米X100米，双向八车道
重要奖项：2003上海市优秀勘察设计二等奖
Location: Pudong New District, Shanghai Municipality **Dates:** 1997
Scale: Two-way road with 8 lanes, 4.4kmX100m
Major award: Second-class Prize of Shanghai for Outstanding Exploration and Design in 2003

101 市政 MUNICIPAL

浦东新区罗山路延长线（外环线—龙东大道）
Extended Part of Luoshan Road
(from the Outer Ring to Longdong Avenue)
in Pudong New District

项目地点： 上海市浦东新区
项目设计时间： 1999年
重要奖项： 2002上海市优秀勘察设计三等奖
项目规模： 3.5千米X80米，双向8车道含互通立交一座
Location: Pudong New District, Shanghai Municipality
Dates: 1999
Scale: 3.5kmX80m, Including a Two-way Road with 8 Lanes and an Interchangeable Flyover
Major award: Third-class Prize of Shanghai for Outstanding Exploration and Design in 2002

102 市政
MUNICIPAL

五洲大道(翔殷路越江隧道—外环线)新建工程
New Construction Project of Wuzhou Avenue
(from Xiangyin Road Crossriver Tunnel to the Outer Ring)

项目地点: 上海市浦东新区
项目设计时间: 2004年
项目规模: 3.3千米(含下立交、大型互通立交各一座)
重要奖项: 上海市优秀勘察设计一等奖
　　　　　2008全国优秀工程勘察设计行业奖
　　　　　市政公用工程三等奖
　　　　　2009国家优质工程银奖

Location: Pudong New District, Shanghai Municipality
Dates: 2004
Scale: 3.3km, Including a Downlink Flyover and a Large-scale Interchangeable Flyover
Major award: First-class Prize of Shanghai for Outstanding Exploration and Design
　　　　　　National Award for Outstanding Exploration and Design in 2008
　　　　　　Third-class Prize for Outstanding Municipal Facility Construction
　　　　　　National Silver Award for Outstanding Construction in 2009

市政 MUNICIPAL

中环线浦东段(上中路越江隧道—申江路)新建工程
New Construction of the Middle Ring in Pudong New District (from Shangzhong Road Crossriver Tunnel to Shenjiang Road)

项目地点： 上海市浦东新区　**项目设计时间：** 2007年　**项目规模：** 2千米×70米（主线长1.5千米占地约363.2亩，枢纽型全互通立交）
重要奖项： 2011上海市优秀工程设计一等奖
　　　　　　2011上海市优秀工程咨询成果二等奖

Location: Pudong New District, Shanghai Municipality　**Dates:** 2007　**Scale:** 2kmX70m, Main line is 1.5 km, Covering 363.2 Mu, and it is a Totally Interchangeable Flyover Hub.
Major award: First-class Prize of Shanghai for Outstanding Construction and Design in 2011
　　　　　　　Second-class Prize of Shanghai for Outstanding Construction Project Consulting in 2011

106 市政
MUNICIPAL

107 市政 MUNICIPAL

贵阳市中坝路改扩建工程
Renovation and Extension Project of Zhongba Road in Guiyang

项目地点: 贵州省贵阳市 **项目设计时间:** 2009年
项目规模: 桥梁长度1435米,桥梁面积54389平方米
Location: Guiyang, Guizhou Province **Dates:** 2009
Scale: Total Length of the Bridge is 1435 m, and Total Area of the Bridge is 54389 sq.m.

金海路(日新路—川南奉公路)新建工程
New Construction of Jinhai Road (Rixin Road to Chuan Nanfeng Highway)

项目地点: 上海市浦东新区
项目设计时间: 2003年
项目规模: 2.83千米X50米,含229米浦东运河桥一座
Location: Pudong New District, Shanghai Municipality
Dates: 2003
Scale: 2.83kmX50m, Including One Canal Bridge with 229 Meters Long in Pudong

周邓公路(申江路—南六公路)新建工程
New Construction Project of Zhoudeng Road (from Shenjiang Road to Nanliu Road)

项目地点：上海市浦东新区　**项目设计时间：**2012年　**项目规模：**5.67千米×45米，双向6车道
Location: Pudong New District, Shanghai Municipality　**Dates:** 2012　**Scale:** 5.67kmX45m, Two-way Road with 6 Lanes

113 市政 MUNICIPAL

杨高路(成山路—外高桥5号门)改建工程
Renovation Project of Yanggao Road
(from Chengshan Road to Waigaoqiao Gate No.5)

项目地点： 上海市浦东新区　**项目设计时间：** 2004年
项目规模： 8.9千米X50米，按双向8快2慢拓宽
Location: Pudong New District, Shanghai Municipality　**Dates:** 2004
Scale: 8.9kmX50m, Bordened into the Two-way Toad with 8 Express Lanes and 2 Slow Lanes

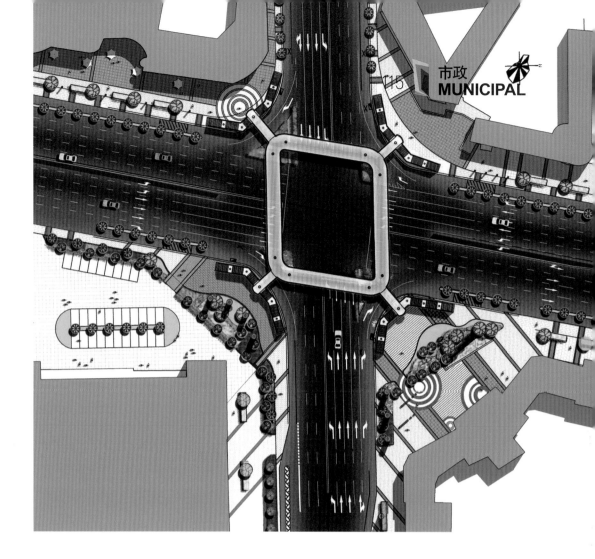

山东省东营市西四路、济南路综合改造工程
Comprehensive Renovation Project of Xisi Road and Jinan Road in Dongying, Shandong Province

项目地点：山东省东营市西城区　**项目设计时间：**2011年　**项目规模：**全长7.6千米X50米
Location: Western City Zone of Dongying, Shandong Province　**Dates:** 2011　**Scale:** Total length 7.6 kmX 50 m

浙江省德清县北湖街延伸工程
Extension Construction Project of Beihu Avenue in Deqing County, Zhejiang Province

项目地点: 浙江省德清县 **项目设计时间:** 2003年
项目规模: 全长4.1千米,红线宽70~75米,双向8车道
重要奖项: 2006浙江省建设工程钱江杯奖(优质工程)
　　　　　　上海市优秀勘察设计三等奖

Location: Deqing County, Zhejiang Province **Dates:** 2003
Scale: Total Length: 4.1km, Width between Red Lines: 70~75 m, Two-way Road with 8 Lanes
Major award: Qianjiang Cup of Zhejiang for Outstanding Construction in 2006
　　　　　　　　Third-class Prize of Shanghai for Outstanding Exploration and Design

江苏省通州市"两路一桥"工程
Double-roads-and-one-bridge Project of Tongzhou, Jiangsu Province

项目地点: 江苏省通州市金沙镇　**项目设计时间:** 2006年　**项目规模:** 全长4.3千米X(34~48) 米
Location: Jinsha Town, Tongzhou, Jiangsu Province　**Dates:** 2006　**Scale:** Total length 4.3kmX(34~48)m

市政 MUNICIPAL

青浦大道（318国道—五浦路）新建工程
New Construction Project of Qingpu Avenue (from the National Highway 318 to Wupu Road)

项目地点：上海市青浦区　**项目设计时间：**2011年　**项目规模：**3千米
Location: Qingpu District, Shanghai Municipality　**Dates:** 2011　**Scale:** 3km

市政
MUNICIPAL

浦建路（杨高南路—龙阳路）、沪南路（前程路—区界）改建工程
Reconstruction Project of Pujian Road (from South Yanggao Road to Longyang Road) and Hunan Road (from Qiancheng Road to the District Border)

项目地点： 上海市浦东新区　**项目设计时间：** 2007年
项目规模： 全长592千米X50米，浦建路由双向4快2慢拓宽为双向8快2慢，沪南路由双向4快2慢拓宽为双向6快2慢
Location: Pudong New District, Shanghai Municipality　**Dates:** 2007
Scale: Total length 592 km x 50 m, Pujian Road is Bordened From the Two-way Road with 4 Express Lanes and 2 Slow Lanes to the Road with 8 Express Lanes and 2 Slow Lanes, Hunan Road is Bordened from the Two-way Road with 4 Express Lanes and 2 Slow Lanes to the Road with 6 Express Lanes and 2 Slow Lanes

市政 MUNICIPAL

东方路改扩建工程
Reconstruction and Expansion Construction Project of Dongfang Road

项目地点： 上海市浦东新区　**项目设计时间：** 2003年　**项目规模：** 4.2千米双向8车道　**重要奖项：** 上海市优秀勘察设计三等奖
Location: Pudong New District, Shanghai　**Dates:** 2003　**Scale:** 4.2km, Two-way with 8 Lanes
Major award: Third-class Prize of Shanghai for Outstanding Exploration and Design

125 市政 MUNICIPAL

市政
MUNICIPAL

莎车县城南新区古勒巴格路、站前路(南环路)道路工程
Construction Project of Gulebage Road and Zhanqian Road (the South Ring) in Nanxin District of Shache Town

项目地点: 新疆维吾尔自治区喀什市莎车县 **项目设计时间:** 2011年 **项目规模:** 道路长度5.55千米
Location: Shache County, Kashi, Xinjiang Uygur Autonomous Region **Dates:** 2011 **Scale:** 5.55km

127 市政 MUNICIPAL

市政 MUNICIPAL

合肥工业园区道路
Road Construction Project of Hefei Industrial Park

项目地点: 安徽省合肥市 **项目设计时间:** 2006年 **项目规模:** 20千米X36米
Location: Hefei, Anhui Province **Dates:** 2006 **Scale:** 20kmX36m

外高桥新市镇配套工程
Infrastructure Construction Project of the New Town in Waigaoqiao District

项目地点: 上海市浦东新区 **项目设计时间:** 2007年 **项目规模:** 4千米
重要奖项: 2009上海市优秀工程咨询成果二等奖
Location: Pudong New District, Shanghai Municipality **Dates:** 2007 **Scale:** 4km
Major award: Second-class Prize of Shanghai for Outstanding Construction Project Consulting in 2009

133 | 市政 MUNICIPAL

长清路（成山路—杨思路）扩建工程
Extension Project of Changqing Road (from Chengshan Road to Yangsi Road)

项目地点： 上海市浦东新区　**项目设计时间：** 2008年　**项目规模：** 2.3千米X(32~40)米
重要奖项： 2010上海市优秀工程咨询成果三等奖
Location: Pudong New District, Shanghai Municipality　**Dates:** 2008　**Scale:** 2.3kmX(32~40) m
Major award: Third-class Prize of Shanghai for Outstanding Construction Project Consulting in 2010

135 市政 MUNICIPAL

苏州河武宁路桥景观改造工程
Landscape Renovation Project of Wuning Road Bridge over Suzhou River

项目地点：上海市普陀区　**项目设计时间：**2008年　**项目规模：**22000平方米
Location: Putuo District, Shanghai Municipality　**Dates:** 2008　**Scale:** 22000 sq.m.

市政
MUNICIPAL

苏州河宝成桥景观改造工程
Landscape Renovation Project
of Baocheng Bridge over Suzhou River

项目地点：上海市普陀区　**项目设计时间：**2009年　**项目规模：**300平方米
Location: Putuo District, Shanghai Municipality　**Dates:** 2009　**Scale:** 300 sq.m.

市政
MUNICIPAL

139 市政 MUNICIPAL

浦东软件园陆家嘴分园步行桥工程
Construction Project of the Pedestrain Bridges in Lujiazui Subzone of Pudong Software Industrial Park

项目地点：上海市浦东新区　**项目设计时间：**2006年　**项目规模：**414平方米
Location: Pudong New District, Shanghai Municipality　**Dates:** 2006　**Scale:** 414 sq.m.

市政
MUNICIPAL

山东省东营市东三路生态廊道桥梁工程
Construction Project of the Bridges in the Ecological Corridor along Dongsan Road in Dongying, Shandong Province

项目地点: 山东省东营市 **项目设计时间:** 2012年 **项目规模:** 69500平方米
Location: Dongying, Shandong Province **Dates:** 2012 **Scale:** 69500 sq.m.

142 市政 MUNICIPAL

吕梁市离石区东城新区东川河桥梁工程
Construction Project of the Bridge over Dongchuan River in the Eastern New Zone of Lishi District of Lvliang

项目地点: 山西省吕梁市离石区　**项目设计时间:** 2009年　**项目规模:** 15600平方米
Location: Lishi District, Lvliang, Shanxi Province　**Dates:** 2009　**Scale:** 15600 sq.m.

143 市政 MUNICIPAL

144 市政
MUNICIPAL

市政
MUNICIPAL

曹路镇第二期自然村落污水纳管工程
The Second Phase Sewage-into-sewer Project of the Existing Villages in Caolu Town

项目地点: 上海市浦东新区曹路镇 **项目设计时间:** 2009年 **项目规模:** 管长约161千米
重要奖项: 2010上海市优秀工程咨询成果二等奖
Location: Caolu Town, Pudong New District, Shanghai Municipality **Dates:** 2009 **Scale:** length 161km
Major award: Second-class Prize of Shanghai for Outstanding Construction Project Consulting in 2010

风景园林　　**LANDSCAPE ARCHITECTURE**

世博公园A块(亩中山水)景观工程
Landscape Project of Zone A of the Expo Park: Mountains and Rivers in Plot A

项目地点： 上海市浦东新区 **项目设计时间：** 2009年 **项目规模：** 20000平方米
重要奖项： 2011上海市优秀工程设计一等奖
第一届优秀风景园林规划设计二等奖

Location: Pudong New District, Shanghai Municipality **Dates:** 2009 **Scale:** 20000 sq.m.
Major award: First-class Prize of Shanghai for Outstanding Construction and Design in 2011
Second-class Prize for Outstanding Landscape Planning and Design in the First National Competition

上海滨江森林公园二期工程
The Second Phase Development Project of Shanghai Waterfront Forest Park

项目地点：上海市浦东新区　**项目设计时间：**2012年　**项目规模：**1190000平方米
重要奖项：2012上海市优秀工程咨询成果二等奖
（与上海市园林设计院联合设计）
Location: Pudong New District, Shanghai Municipality　**Dates:** 2012　**Scale:** 1190000 sq.m.
Major award: Second-class Prize of Shanghai for Outstanding Construction Project Consulting in 2012
(To Design With Shanghai Landscape Architecture Design Institute)

风景园林
LANDSCAPE ARCHITECTURE

风景园林
LANDSCAPE ARCHITECTURE

江苏省徐州市行政中心
市民中心广场绿化环境景观工程
Virescence and Landscape Project of the Plaza of Administration and Civic Center in Xuzhou, Jiangsu Province

项目地点：江苏省徐州市　**项目设计时间：**2006年　**项目规模：**240000平方米
重要奖项：2009上海市优秀工程设计二等奖
Location: Xuzhou, Jiangsu Province　**Dates:** 2006　**Scale:** 240000 sq.m.
Major award: Second-class Prize of Shanghai for Outstanding Construction and Design in 2009

155 风景园林
LANDSCAPE ARCHITECTURE

海门市民公园规划设计
Urban Design of Haimen Civic Park

项目地点: 江苏海门市　**项目设计时间:** 2011年　**项目规模:** 680000平方米
Location: Haimen, Jiangsu Province　**Dates:** 2011　**Scale:** 680000 sq. m

山东东营西湖景观工程
Landscape Construction of the West Lake of Dongying in Shandong

项目地点：山东省东营市　**项目设计时间**：2012年　**项目规模**：272000平方米
Location: Dongying, Shandong Province　**Dates:** 2012　**Scale:** 272000 sq.m.

梅园公园改造工程
Renovation Project of Meiyuan Park

项目地点： 上海市浦东新区　**项目设计时间：** 2008年　**项目规模：** 18000平方米
重要奖项： 2010上海市优秀工程咨询成果三等奖
　　　　　2011上海市优秀工程设计三等奖

Location: Pudong New District, Shanghai Municipality　**Dates:** 2008　**Scale:** 18000 sq.m.
Major award: Third-class Prize of Shanghai for Outstanding Construction Project Consulting in 2010
　　　　　　　Third-class Prize of Shanghai for Outstanding Construction and Design in 2011

风景园林
LANDSCAPE ARCHITECTURE

长风工业园地块公共绿地建设项目
Construction Project of the Public Green Spaces in Changfeng Industrial Park

项目地点： 上海市普陀区　**项目设计时间：** 2009年　**项目规模：** 110000平方米
Location: Putuo District, Shanghai Municipality　**Dates:** 2009　**Scale:** 110000 sq.m.

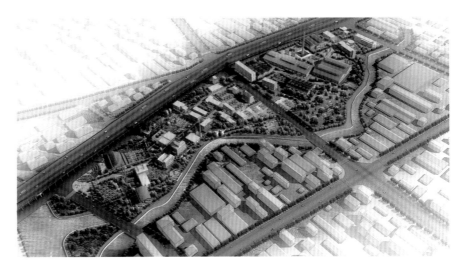

风景园林
LANDSCAPE ARCHITECTURE

张衡公园
Zhangheng Park

项目地点: 上海市浦东新区　**项目设计时间:** 2008年　**项目规模:** 74200平方米
重要奖项: 2010上海市优秀工程咨询成果奖二等奖
　　　　　　 2013上海市优秀工程设计三等奖

Location: Pudong New District, Shanghai Municipality　**Dates:** 2008　**Scale:** 74200 sq.m.
Major award: Second-class Prize of Shanghai for Outstanding Construction Project Consulting in 2010
　　　　　　　　Third-prize of Shanghai for Outstanding Construction and Design in 2013

江西鹰潭市滨江公园二期景观工程
The Second Phase Landscape Project of Yingtan Waterfront Park in Jiangxi Province

项目地点: 江西省鹰潭市 **项目设计时间:** 2010年 **项目规模:** 250000平方米
重要奖项: 2012上海市优秀工程咨询成果三等奖
Location: Yingtan, Jiangxi Province **Dates:** 2010 **Scale:** 250000 sq.m.
Major award: Third-class Prize of Shanghai for Outstanding Construction Project Consulting in 2012

风景园林
LANDSCAPE ARCHITECTURE

风景园林
LANDSCAPE ARCHITECTURE

海门市圩角河两侧绿化景观建设项目
Landscape Design of the Riverfront of Weijiao River in Haimen

项目地点： 江苏省海门市　**项目设计时间：** 2013年　**项目规模：** 960000平方米
Location: Haimen, Jiangsu Province　**Dates:** 2013　**Scale:** 960000 sq.m.

风景园林
LANDSCAPE ARCHITECTURE

外滩滨江绿地景观工程
Landscape Project of the Waterfront Green Spaces of the Bund

项目地点：上海市浦东新区　**项目设计时间：**2003年　**项目规模：**70000平方米
Location: Pudong New District, Shanghai Municipality　**Dates:** 2003　**Scale:** 70000 sq.m.

风景园林
LANDSCAPE ARCHITECTURE

东营市东三路河生态廊道（德州路—东营河）景观工程
Landscape Construction Project of the Ecological Corridor along Dongshanlu River in Dongying (from Dezhou Road to Dongying River)

项目地点： 山东省东营市　**项目设计时间：** 2012年　**项目规模：** 536600平方米
Location: Dongying, Shandong Province　**Dates:** 2012　**Scale:** 536600 sq.m.

171 | 风景园林
LANDSCAPE ARCHITECTURE

风景园林
LANDSCAPE ARCHITECTURE

风景园林
LANDSCAPE ARCHITECTURE

三林世博家园公共绿地建设工程
Construction Project of the Public Green Space in Sanlin Expo Home

项目地点: 上海市浦东新区　**项目设计时间:** 2005年　**项目规模:** 140000平方米
重要奖项: 上海市优秀勘察设计三等奖
Location: Pudong New District, Shanghai Municipality　**Dates:** 2005　**Scale:** 140000 sq.m.
Major award: Third-class Prize of Shanghai for Outstanding Exploration and Design

风景园林
LANDSCAPE ARCHITECTURE

风景园林
LANDSCAPE ARCHITECTURE

保利湖畔阳光苑景观工程
Landscape Project of the Lakefront Sunlight Garden Developed by Baoli Real Estate

项目地点: 上海市嘉定区　**项目设计时间:** 2009年　**项目规模:** 97000平方米
Location: Jiading District, Shanghai Municipality　**Dates:** 2009　**Scale:** 97000 sq.m.

风景园林
LANDSCAPE ARCHITECTURE

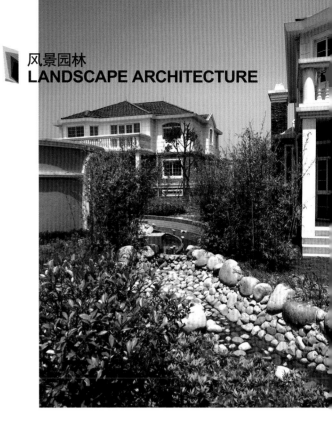

南通东郊庄园
Eastern Suburb Manor in Nantong

项目地点: 江苏省南通市　**项目设计时间:** 2005年　**项目规模:** 124000平方米
重要奖项: 2008上海市优秀住宅设计三等奖
Location: Nantong, Jiangsu Province　**Dates:** 2005　**Scale:** 124000 sq.m.
Major award: Third-class Prize of Shanghai for Outstanding House Design in 2008

湖南电力科技园景观工程
Landscape Project of the Electrical Science and Technology Park in Hunan

项目地点: 湖南省长沙市　**项目设计时间:** 2005年　**项目规模:** 137300平方米
重要奖项: 上海市优秀勘察设计三等奖
Location: Changsha, Hunan Province　**Dates:** 2005　**Scale:** 137300 sq.m.
Major award: Third-class Prize of Shanghai for Outstanding Exploration and Design

风景园林
LANDSCAPE ARCHITECTURE

风景园林
LANDSCAPE ARCHITECTURE

永达大厦景观绿化工程
Landscape and Virescence Project of Yongda Building

项目地点: 上海市浦东新区　**项目设计时间:** 2006年　**项目规模:** 15000平方米
Location: Pudong New District, Shanghai Municipality　**Dates:** 2006　**Scale:** 15000 sq.m.

大渡口H21—1经济适用房项目绿化景观工程设计
Virescence and Landscape Project of Affordable Housing in Zone H21-1 of Dadukou District

项目地点:重庆市大渡口区　项目设计时间:2010年　项目规模:50000平方米
Location: Dadukou District, Chongqing Municipality　Dates: 2010　Scale: 50000 sq.m.

风景园林
LANDSCAPE ARCHITECTURE

江苏省泰州市济川医药工业园(一期)景观工程
The First Phase Landscape Project of Jichuan Medicine Industry Park in Taizhou, Jiangsu Province

项目地点: 江苏省泰兴市　**项目设计时间:** 2006年　**项目规模:** 200000平方米
Location: Taixing, Jiangsu Province　**Dates:** 2006　**Scale:** 200000 sq.m.

风景园林
LANDSCAPE ARCHITECTURE

浦东南路道路景观工程
Landscape Project of South Pudong Road

项目地点： 上海市浦东新区　**项目设计时间：** 2009年　**项目规模：** 600000平方米
重要奖项： 2011上海市优秀工程设计二等奖
　　　　　　第一届优秀风景园林规划设计三等奖

Location: Pudong New District, Shanghai Municipality　**Dates:** 2009　**Scale:** 600000 sq.m.
Major award: Second-class Prize of Shanghai for Outstanding Construction and Design in 2011
　　　　　　　　Third-class Prize in the First Landscape Architectural Design and Planning Contest

中环线浦东段新建及两侧绿化工程
New Construction and Virescence Project of the Middel Ring in Pudong New District

项目地点：上海市浦东新区　**项目设计：**2008年　**项目规模：**1290000平方米
重要奖项：2010上海市优秀工程咨询成果三等奖
Location: Pudong New District, Shanghai Municipality　**Dates:** 2008　**Scale:** 1290000 sq.m.
Major award: Third-class Prize of Shanghai for Outstanding Construction Project Consulting in 2010

临港新城申港大道景观工程
Landscape Project of Shengang Avenue in Lingang New Town

项目地点: 上海市浦东新区 **项目设计时间:** 2004年 **项目规模:** 380000平方米
重要奖项: 上海市优秀勘察设计一等奖
Location: Pudong New District, Shanghai Municipality **Dates:** 2004 **Scale:** 380000 sq.m.
Major award: First-class Prize of Shanghai for Outstanding Exploration and Design

风景园林
LANDSCAPE ARCHITECTURE

东方路景观
Landscape of Dongfang Road

项目地点: 上海市浦东新区　**项目设计时间:** 2005年　**项目规模:** 70000平方米
重要奖项: 上海市优秀工程设计二等奖
Location: Pudong New District, Shanghai Municipality　**Dates:** 2005　**Scale:** 70000 sq.m.
Major award: Second-class Prize of Shanghai for Outstanding Construction and Design

江苏海门北京路景观工程
Landscape Project of Beijing Road in Haimen, Jiangsu Province

项目地点: 江苏省海门市 **项目设计时间:** 2008年 **项目规模:** 320000平方米
Location: Haimen, Jiangsu Province **Dates:** 2008 **Scale:** 320000 sq.m.

风景园林
LANDSCAPE ARCHITECTURE

五洲大道(浦东北路—外环线)道路绿化工程
Road Virescence Project of Wuzhou Avenue (from North Pudong Road to the Outer Ring)

项目地点: 上海市浦东新区　**项目设计时间:** 2005年　**项目规模:** 422000平方米
重要奖项: 上海市优秀勘察设计一等奖
Location: Pudong New District, Shanghai Municipality　**Dates:** 2005　**Scale:** 422000 sq.m.
Major award: First-class Prize of Shanghai for Outstanding Exploration and Design

室内装饰　　**INTERIOR**

上海五角场万达商业广场地下一层购物中心室内装饰工程
Interior Decoration Project of the Underground First Floor Shopping Center in Wanda Plaza in Wujiaochang District, Shanghai

项目地点: 上海市杨浦区 **项目设计时间:** 2006年 **项目规模:** 14000平方米 **重要奖项:** 2010上海市建筑装饰奖
Location: Yangpu District, Shanghai Municipality **Dates:** 2006 **Scale:** 14000 sq.m. **Major award:** Architectural Decoration Prize of Shanghai in 2010

203 室内装饰 INTERIOR

上海周浦万达广场商业步行街室内装修设计
Interior Decoration Design of Commercial Walking Street of Wanda Plaza in Zhoupu District, Shanghai

项目地点: 上海市浦东新区周浦镇　**项目设计时间:** 2009年　**项目规模:** 18000平方米
Location: Zhoufu Town, Pudong New District, Shanghai Municipality　**Dates:** 2009　**Scale:** 18000 sq.m.

上海宝山万达广场室内商业步行街公共空间室内设计
Interior Public Space Design of Commercial Walking Street of Wanda Plaza in Baoshan District, Shanghai

项目地点：上海宝山区　**项目设计时间：**2011年　**项目规模：**18300平方米
Location: Baoshan District, Shanghai Municipality　**Dates:** 2011　**Scale:** 18300 sq.m.

南京建邺万达商业广场商业步行街室内设计
Interior Design of the Commercial Walking Street of Jianye Wanda Plaza in Nanjing

项目地点: 江苏省南京市　**项目设计时间:** 2009年　**项目规模:** 21000平方米
Location: Nanjing, Jiangsu Province　**Dates:** 2009　**Scale:** 21000 sq.m.

绍兴柯桥万达广场室内商业步行街室内设计
Interior Design of the Commercial Walking Street of Keqiao Wanda Plaza in Shaoxing

项目地点: 浙江省绍兴市　**项目设计时间:** 2010年　**项目规模:** 17500平方米
Location: Shaoxing, Zhejiang Province　**Dates:** 2010　**Scale:** 17500 sq.m.

室内装饰
INTERIOR

长春宽城万达购物中心步行街公共空间室内设计
Interior Public Space Design of the Walking Street of Wanda Shopping Center in Kuancheng, Changchun

项目地点： 长春市宽城区　**项目设计时间：** 2012年　**项目规模：** 15000平方米
Location: Kuancheng, Changchun　**Dates:** 2012　**Scale:** 15000 sq.m.

室内装饰 INTERIOR

南通中央商务区A-04地块项目室内设计
Interior Design of A-04 Zone in Central Business District of Nantong

项目地点: 江苏省南通市 **项目设计时间:** 2011年 **项目规模:** 25000平方米
Location: Nantong, Jiangsu Province **Dates:** 2011 **Scale:** 25000 sq.m.

213 室内装饰 INTERIOR

盐城中南·世纪城（2A地块）商场公共部位室内设计
Interior Public Space Design of Zhongnan – Century City Market in Yancheng (2A Zone)

项目地点： 江苏省盐城市　**项目设计时间：** 2013年　**项目规模：** 29255.7平方米
Location: Yancheng, Jiangsu Province　**Dates:** 2013　**Scale:** 29255.7 sq.m.

216 室内装饰
INTERIOR

长沙奥克斯广场步行街公共空间室内设计
Interior Public Space Design
of Walking Street of AUX Plaza in Changsha

项目地点: 湖南省长沙市大河西先导区　**项目设计时间:** 2012年　**项目规模:** 9000平方米
Location: Xiandao District, Dahexi, Changsha, Hunan Province　**Dates:** 2012　**Scale:** 9000 sq.m.

上海嘉定方舟广场步行街公共空间设计
Interior Public Space Design
of Walking Street of Fangzhou Plaza in Jiading, Shanghai

项目地点: 上海市嘉定区　**项目设计时间:** 2011年　**项目规模:** 15000平方米
Location: Jiading, Shanghai Municipality　**Dates:** 2011　**Scale:** 15000 sq.m.

南通五洲国际广场公共空间装修设计
Interior Public Space Decoration Design of Wuzhou International Plaza of Nantong

项目地点：江苏省南通市　**项目设计时间：**2013年　**项目规模：**27309.7平方米
Location: Nantong, Jiangsu Province　**Dates:** 2013　**Scale:** 27309.7 sq.m.

室内装饰
INTERIOR

上海爱法小天地酒店式公寓(吉瑞商务酒店)室内设计
Interior Decoration Design of "Shanghai Aifa: Little World" Service Apartment (J-real Residence Suites)

项目地点: 上海市浦东新区　**项目设计时间:** 2007年　**项目规模:** 13000平方米
重要奖项: 2010上海市建筑装饰奖
Location: Pudong New District, Shanghai Municipality　**Dates:** 2007　**Scale:** 13000 sq.m.
Major award: Architectural Decoration Award of Shanghai, 2010

新疆宏泰房产开发旅游宾馆室内装修设计
Interior Decoration Design of the Tourism Hotel Developed by Hongtai Real Estate in Xinjiang

项目地点: 新疆维吾尔自治区阿勒泰地区　**项目设计时间:** 2012年　**项目规模:** 22000平方米
Location: Altay, Xinjiang Uygur Autonomous Region　**Dates:** 2012　**Scale:** 22000 sq.m.

上海张江泰豪智能电气有限公司办公楼室内设计
Office building Interior Decoration Design of Shanghai Zhangjiang Tellhow Intelligent Engineering Co., Ltd.

项目地点：上海市浦东新区　**项目设计时间：**2007年　**项目规模：**3000平方米
Location: Pudong New District, Shanghai Municipality　**Dates:** 2007　**Scale:** 3000 sq.m.

上海奉贤绿庭国际中心写字楼室内装修设计
Interior Decoration Design of Office Buildings of Lvting International Center in Fengxian District, Shanghai

项目地点: 上海市奉贤区　**项目设计时间:** 2012年　**项目规模:** 8782平方米
Location: Fengxian District, Shanghai Municipality　**Dates:** 2012　**Scale:** 8782 sq.m.

上海爱法奥朗别墅二期样板房设计
Model House Design of the Second Phase Development of Aifa: Orang Villa in Shanghai

项目地点: 上海市航头镇 **项目设计时间:** 2008年 **项目规模:** 800平方米
重要奖项: 2010上海市建筑装饰奖
Location: Hangtou Town, Shanghai Municipality **Dates:** 2008 **Scale:** 800 sq.m.
Major award: Architectural Decoration Prize of Shanghai in 2010

合肥吉瑞泰盛国际生活广场售楼处及样板房室内设计
Sales Office and Model House Interior Decoration Design of Jerry Tightsen International Living Mall in Hefei

项目地点： 安徽省合肥市　**项目设计时间：** 2008年　**项目规模：** 1768平方米
重要奖项： 2010上海市建筑装饰奖
Location: Hefei, Anhui Province　**Dates:** 2008　**Scale:** 1768 sq.m.
Major award: Architectural Decoration Award of Shanghai 2010

235 室内装饰 INTERIOR

上海松江绿庭广场高层公寓、会所、公寓式办公楼精装修设计
Refined Decoration Design of Apartments, Clubs and Service Apartments of Lvting Plaza in Songjiang District, Shanghai

项目地点: 上海市松江区　**项目设计时间:** 2010年　**项目规模:** 3200平方米
Location: Songjiang District, Shanghai Municipality　**Dates:** 2010　**Scale:** 3200 sq.m.

上海绿庭尚城住宅精装修户型室内设计
Interior Refined Decoration Design of Shanghai Lvting Residential District

项目地点： 上海市松江区　**项目设计时间：** 2011年　**项目规模：** 900平方米
重要奖项： 2012上海第十一届建筑装饰设计大赛二等奖
Location: Songjiang District, Shanghai　**Dates:** 2012　**Scale:** 900 sq.m.
Major award: Second-class prize of the 11th Shanghai architectural design competition in 2012

城市规划　　**URBAN PLANNING**

240 城市规划
URBAN PLANNING

中国2010年上海世博会园区
浦东场地公共空间规划设计
Public Space Planning of Pudong Zone for Shanghai Expo Region, 2010

项目地点: 上海市浦东新区　**项目设计时间:** 2007年　**项目规模:** 380公顷
重要奖项: 2009全国优秀规划设计三等奖
　　　　　　第一届优秀风景园林规划设计二等奖

Location: Pudong New District, Shanghai Municipality　**Dates:** 2007　**Scale:** 380 ha.
Major award: National Third-class Prize for Outstanding Urban Planning in 2009
　　　　　　　Second-class Prize in the First Landscape Architectural Design and Planning Contest

城市规划
URBAN PLANNING

山西吕梁北川河片区修建性详细规划
Site Detailed Planning of Hepian Area of Beichuan, Lvliang, Shanxi

项目地点：山西省吕梁市　**项目设计时间：**2011年　**项目规模：**161公顷
重要奖项：2011全国人居经典建筑规划设计方案竞赛建筑、科技双金奖
Location: Lvliang, Shanxi Province　**Dates:** 2011　**Scale:** 161 ha.
Major award: National Gold Award for Outstanding Architectural Design and Technology in Residence in 2011

243 城市规划
URBAN PLANNING

洪先路金融文化街概念性规划设计及建筑方案设计
Conceptual Planning and Architectural Design of Financial and Cultural Streetscape along Hongxian Road

项目地点: 江西省吉水县　**项目设计时间:** 2012年　**项目规模:** 8公顷
Location: Jishui County, Jiangxi Province　**Dates:** 2012　**Scale:** 8 ha.

城市规划
URBAN PLANNING

浙江省诸暨市行政中心概念规划及城市设计
Conceptual Planning and Urban Design of Zhuji Civic Center in Zhejiang Province

项目地点: 浙江省诸暨市　**项目设计时间:** 2004年　**项目规模:** 120公顷
Location: Zhuji, Zhejiang Province　**Dates:** 2004　**Scale:** 120 ha.

西宁市湟中县职业教育多巴新校区修建性详细规划
Site Detailed Planning of Duoba New Campus for Vocational Education in Huangzhong County of Xining

项目地点: 青海省西宁市湟中县　**项目设计时间:** 2012年　**项目规模:** 6公顷
Location: Huangzhong County, Xining, Qinghai Province　**Dates:** 2012　**Scale:** 6 ha.

城市规划
URBAN PLANNING

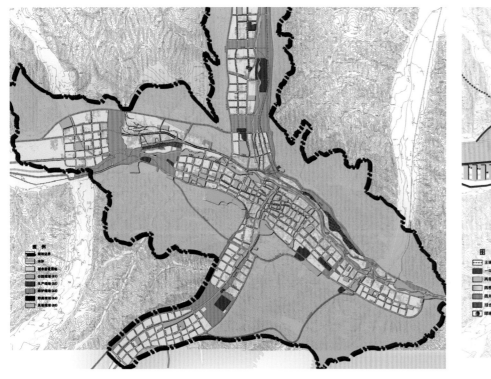
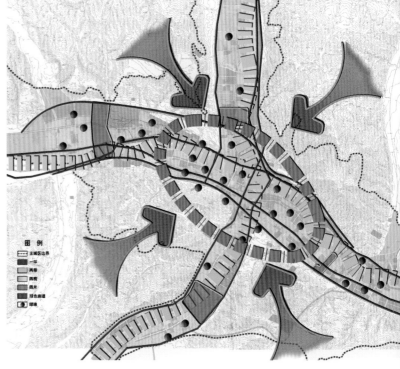

西宁市城市绿地系统规划
Urban Green Space System Planning of Xining

项目地点: 青海省西宁市　**项目设计时间:** 2006年　**项目规模:** 128平方公里
Location: Xining, Qinghai Province　**Dates:** 2006　**Scale:** 128 sq.km.

浦东新区道路建设"十二五"规划
The Twelfth Five-year Plan of Road Construction in Pudong New District

项目地点: 上海浦东新区　**项目设计时间:** 2011年
Location: Pudong New District, Shanghai Municipality　**Dates:** 2011

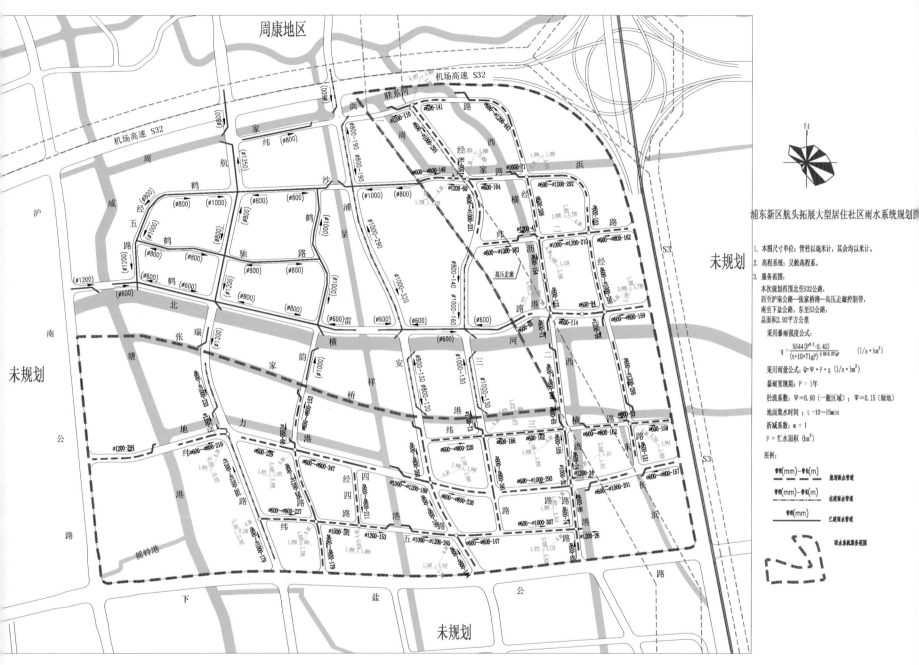

浦东新区航头拓展大型居住社区雨污水系统专业规划
Rainwater and Sewage System
Planning of Hangtou Expanded Major Community in Pudong New District

项目地点: 上海市浦东新区 **项目设计时间:** 2010年 **项目规模:** 292公顷 **重要奖项:** 2012上海市优秀工程咨询成果三等奖
Location: Pudong New District, Shanghai Municipality **Dates:** 2010 **Scale:** 292 ha. **Major award:** Third-class Prize of Shanghai for Outstanding Construction Project Consulting in 2012

上海唐镇新市镇核心区公共空间规划设计导则
Design Guidelines of Public Space Planning in the New Central Region of Tang Town, Shanghai

项目地点: 上海市浦东新区 **项目设计时间:** 2010年 **项目规模:** 420公顷
Location: Pudong New District, Shanghai Municipality **Dates:** 2010 **Scale:** 420 ha.

西宁市城北区小桥东片区控制性详细规划
Regulatory Detailed Plan of Xiaoqiao East Zone in North City District of Xining

项目地点： 青海省西宁市城北区 **项目设计时间：** 2011年 **项目规模：** 250公顷
Location: Northern Zone, Xining, Qinghai Province **Dates:** 2011 **Scale:** 250 ha.

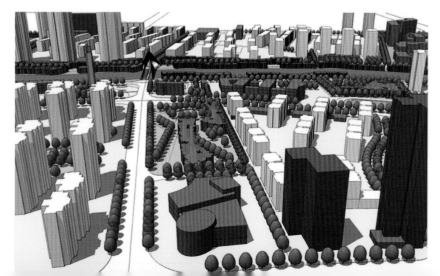

西宁市北川河景观风貌概念规划
Conceptual Planning of Landscape and Scenery along Beichuan River in Xining

项目地点: 青海省西宁市城北区　**项目设计时间:** 2011年　**项目规模:** 900公顷
Location: Northern Zone, Xining, Qinghai Province　**Dates:** 2011　**Scale:** 900 ha.

河道整治　　**WATERWAY DREDGING**

254 河道整治 **WATERWAY DREDGING**

外高桥南块微电子产业基地西区二期水系建设工程
Second Phase Water Course Construction Project of West Zone of Waigaoqiao Southern Microelectronic Industrial Base

项目地点: 上海市浦东新区外高桥 **项目设计时间:** 2012年 **项目规模:** 230000平方米
Location: Waigaoqiao Zone, Pudong New District, Shanghai Municipality **Dates:** 2012 **Scale:** 230000 sq.m.

255 河道整治
WATERWAY DREDGING

浦东北路河(洲海路—椿树浦)河道新建工程
New Construction Project of North Pudong Road Channel (from Zhouhai Road to Chunshupu River)

项目地点: 上海市浦东新区　**项目设计时间:** 2009年　**项目规模:** 0.66千米
Location: Pudong New District, Shanghai Municipality　**Dates:** 2009　**Scale:** 0.66km

河道整治
WATERWAY DREDGING

椿树浦(浦东北路—张杨北路)河道整治工程
Comprehensive Renovation Project of Chunshupu River (from North Pudong Road to North Zhangyang Road)

项目地点： 上海市浦东新区　**项目设计时间：** 2007年　**项目规模：** 1.3千米
Location: Pudong New District, Shanghai Municipality　**Dates:** 2007　**Scale:** 1.3km

257 河道整治 WATERWAY DREDGING

春塘河(川杨河—世博家园)河道综合整治工程
Comprehensive Renovation Project of Chuntang River (from Chuanyang River to Expo Home)

项目地点: 上海市浦东新区 **项目设计时间:** 2007年 **项目规模:** 0.898千米×20米
Location: Pudong New District, Shanghai Municipality **Dates:** 2007 **Scale:** 0.898km × 20m

258 河道整治
WATERWAY DREDGING

中汾泾(川杨河—杨高南路)河道综合整治工程
Comprehensive Renovation Project of Zhongfenjing River (from Chuanyang River to South Yanggao Road)

项目地点： 上海市浦东新区　**项目设计时间：** 2007年　**项目规模：** 0.785千米X26米
Location: Pudong New District, Shanghai Municipality　**Dates:** 2007　**Scale:** 0.785kmX26m

中心河(川杨河—华夏西路)河道综合整治工程
Comprehensive Renovation Project of the Central River (from Chuanyang River to West Huaxia Road)

项目地点:上海市浦东新区 **项目设计时间:**2007年 **项目规模:**1.6千米X20米
Location: Pudong New District, Shanghai Municipality **Dates:** 2007 **Scale:** 1.6km X 20m

河道整治
WATERWAY DREDGING

大寨河(高青路—华夏西路)河道综合整治工程
Comprehensive Renovation Project of Dazhai River (from Gaoqing Road to West Huaxia Road)

项目地点: 上海市浦东新区　**项目设计时间:** 2007年　**项目规模:** 0.627千米×28米
Location: Pudong New District, Shanghai Municipality　**Dates:** 2007　**Scale:** 0.627km × 28m

河道整治
WATERWAY DREDGING

大寨河(川杨河—高青路)河道综合整治工程
Comprehensive Renovation Project of Dazhai River (from Chuanyang River to Gaoqing Road)

项目地点: 上海市浦东新区　**项目设计时间:** 2007年　**项目规模:** 0.785千米X26米
Location: Pudong New District, Shanghai Municipality　**Dates:** 2007　**Scale:** 0.785kmX26m

薛家浜(大寨河—沪南公路)河道综合整治工程
Comprehensive Renovation Project of Xuejiabang River (from Dazhai River to Hunan Road)

项目地点: 上海市浦东新区　**项目设计时间:** 2007年　**项目规模:** 0.596千米×20米
Location: Pudong New District, Shanghai Municipality　**Dates:** 2007　**Scale:** 0.596km × 20m

263 河道整治 | WATERWAY DREDGING

中心河(白莲泾—川杨河)河道综合整治工程
Comprehensive Renovation Project of the Central River (from Bailianjing River to Chuanyang River)

项目地点: 上海市浦东新区 **项目设计时间:** 2007年 **项目规模:** 0.94千米×26米
Location: Pudong New District, Shanghai Municipality **Dates:** 2007 **Scale:** 0.94km × 26m

工程咨询　　**ENGINEERING CONSULTING**

工程咨询
ENGINEERING CONSULTING

2012

- **谢稚柳陈佩秋艺术馆工程可行性研究报告**
Feasibility Study Report of Xie Zhiliu & Chen Peiqiu Art Gallery Construction Project
2012上海市优秀工程咨询成果三等奖
Award: Third-class Prize of Shanghai for Outstanding Construction Project Consulting in 2012

- **黎明资源再利用中心基坑安全性评估报告及围护设计评审方案**
Security Evaluation Report on the Foundation Pit of Liming Resource Recycling Center and Envelope Design Scheme for Assessment
2012上海市优秀工程咨询成果二等奖
Award: Second-class Prize of Shanghai for Outstanding Construction Project Consulting in 2012

- **上海市消防局2010年消防营房维修等项目可研性研究报告**
Feasibility Study Report of Fire Station Renovation Project of Shanghai Public Fire Department in 2010
2012上海市优秀工程咨询成果三等奖
Award: Third-class Prize of Shanghai for Outstanding Construction Project Consulting in 2012

- **临港主城区WNW-C5-2地块初级中学工程可研性研究报告**
Feasibility Study Report of Middle School Project in WNW-C5-2 Region of Lingang Major Urban Area
2012上海市优秀工程咨询成果一等奖
Award: First-class Prize of Shanghai for Outstanding Construction Project Consulting in 2012

- **浦东新区道路养护道班房"十二五"规划研究**
The Twelfth Five-Year Planning of Road Maintenance Stations in Pudong New District (2011-2015)
2012上海市优秀工程咨询成果三等奖
Award: Third-class Prize of Shanghai for Outstanding Construction Project Consulting in 2012

- **浦东新区生态河道建设和管理指导意见**
Guidelines of Construction and Management of Ecological Watercourses in Pudong New District
2012上海市优秀工程咨询成果三等奖
Award: Third-class Prize of Shanghai for Outstanding Construction Project Consulting in 2012

- **2012年浦东新区环保市容局基建项目方案研究**
Schematic Study on the Infrastructure Construction Project of Environmental Protection and Townscape Bureau of Pudong New District in 2010
2012上海市优秀工程咨询成果三等奖
Award: Third-class Prize of Shanghai for Outstanding Construction Project Consulting in 2012

- **淀山湖大道西延伸段工程**
West Extension Project of Dianshan Lake Avenue
2012上海市优秀工程咨询成果三等奖
Award: Third-class Prize of Shanghai for Outstanding Construction Project Consulting in 2012

- **浙江省德清县舞阳街工程**
Wuyang Avenue Construction Project of Deqing County in Zhejiang Province
2012上海市优秀工程咨询成果二等奖
Award: Second-class Prize of Shanghai for Outstanding Construction Project Consulting in 2012

工程咨询
ENGINEERING CONSULTING

2012

- **上海市浦东新区未纳管污染源调查研究**
Investigation Report on the Sewage-without-sewer Pollution Sources in Pudong New District, Shanghai
2012上海市优秀工程咨询成果二等奖
Award: Second-class Prize of Shanghai for Outstanding Construction Project Consulting in 2012

- **浦东新区2010年村庄改造项目—曹路低水压管网改造工程**
Renovation Project of Low Pressure Water Pipes in Chaolu Town: a Village Renovation Project of Pudong New District in 2010
2012上海市优秀工程咨询成果三等奖
Award: Third-class Prize of Shanghai for Outstanding Construction Project Consulting in 2012

- **南六污水支线工程**
Southern Six Sewage Branch Pipeline Construction Project
2012上海市优秀工程咨询成果三等奖
Award: Third-class Prize of Shanghai for Outstanding Construction Project Consulting in 2012

- **浦东新区2011年村庄改造项目—祝桥镇污水治理工程**
Sewage Treatment Project of Zhuqiao Town: a Village Renovation Project of Pudong New District in 2010
2012上海市优秀工程咨询成果三等奖
Award: Third-class Prize of Shanghai for Outstanding Construction Project Consulting in 2012

- **浦东新区航头拓展大型居住社区雨污水系统专业规划**
Rainwater and Sewage System Planning of Hangtou Expanded Major Community in Pudong New District
2012上海市优秀工程咨询成果三等奖
Award: Third-class Prize of Shanghai for Outstanding Construction Project Consulting in 2012

- **滨江森林公园二期工程**
Second Phase Construction of Waterfront Forest Park
2012上海市优秀工程咨询成果二等奖
Award: Second-class Prize of Shanghai for Outstanding Construction Project Consulting in 2012

- **南汇生态专项建设工程**
Nanhui Ecological Construction Project
2012上海市优秀工程咨询成果三等奖
Award: Third-class Prize of Shanghai for Outstanding Construction Project Consulting in 2012

- **曙光绿地建设工程**
Shuguang Virescence Construction Project
2012上海市优秀工程咨询成果三等奖
Award: Third-class Prize of Shanghai for Outstanding Construction Project Consulting in 2012

- **陆家嘴金融城公共空间功能调整完善实施课题研究**
Subject Study on the Adjustment, Improvement and Implementation of Public Space Planning in Lujiazui Financial City
2012上海市优秀工程咨询成果二等奖
Award: Second-class Prize of Shanghai for Outstanding Construction Project Consulting in 2012

工程咨询
ENGINEERING CONSULTING

2012

- **江西鹰潭滨江公园规划设计方案专题研究**
  Subject Study on the Conceptual Planning of Yingtan Waterfront Park in Jiangxi Province
  2012上海市优秀工程咨询成果三等奖
  Award: Third-class Prize of Shanghai for Outstanding Construction Project Consulting in 2012

- **浦东软件园三期A1地块综合商业配套项目建筑节能评估报告**
  Building Energy Performance Evaluation Report of Comprehensive Commercial Facilities in Area A1 of Phase 3 in Pudong Software Industrial Park
  2012上海市优秀工程咨询成果三等奖
  Award: Third-class Prize of Shanghai for Outstanding Construction Project Consulting in 2012

- **上海富士康大厦项目建筑节能评估报告**
  Building Energy Performance Evaluation Report of Shanghai Foxconn Building
  2012上海市优秀工程咨询成果二等奖
  Award: Second-class Prize of Shanghai for Outstanding Construction Project Consulting in 2012

2011

- **罗山路(龙东大道—S20)快速化改建工程**
  Reconstruction Project of Luoshan Express Way (Longdong Avenue-S20)
  2011上海市优秀工程咨询成果二等奖
  Award: Second-class Prize of Shanghai for Outstanding Construction Project Consulting in 2011

- **中环线浦东段(军工路越江隧道—高科中路)新建工程**
  New Construction Project of Middle Ring in Pudong (from Jungong Road Crossriver Tunnel to Gaoke Middle Road)
  2011上海市优秀工程咨询成果二等奖
  Award: Second-class Prize of Shanghai for Outstanding Construction Project Consulting in 2011

- **浦东新区2011年村庄改造工程—合庆镇低水压管网改造治工程**
  Renovation Project of Low Pressure Water Pipe System of Heqing Town: a Village Renovation Project of Pudong New District in 2010
  2011上海市优秀工程咨询成果二等奖
  Award: Second-class Prize of Shanghai for Outstanding Construction Project Consulting in 2011

- **曹路文化中心可行性研究报告**
  Feasibility Study Report of Caolu Cultural Center
  2011上海市优秀工程咨询成果三等奖
  Award: Third-class Prize of Shanghai for Outstanding Construction Project Consulting in 2011

- **上海张江集电港B区4-9幼托项目申请报告**
  Application Report of 4-9 Kindergarten Project in Zone B of Zhangjiang Jidiangang in Shanghai
  2011上海市优秀工程咨询成果三等奖
  Award: Third-class Prize of Shanghai for Outstanding Construction Project Consulting in 2011

- **上海迪斯尼项目配套项目**
  Supporting Project of Shanghai Disneyland
  2011上海市优秀工程咨询成果三等奖
  Award: Third-class Prize of Shanghai for Outstanding Construction Project Consulting in 2011

工程咨询
ENGINEERING CONSULTING

2011

- **东靖路（申江路—华东路）新建工程**
  New Construction Project of Dongjing Road (from Shenjiang Road to Huadong Road)
  2011上海市优秀工程咨询成果三等奖
  Award: Third-class Prize of Shanghai for Outstanding Construction Project Consulting in 2011

- **浦东新区2010年村庄改造工程—合庆镇污水治理工程**
  Sewage Treatment Project of Heqing Town: a Village Renovation Project of Pudong New District in 2010
  2011上海市优秀工程咨询成果三等奖
  Award: Third-class Prize of Shanghai for Outstanding Construction Project Consulting in 2011

- **长泰国际广场建筑节能评估报告**
  Building Energy Performance Evaluation Report of Changtai International Plaza
  2011上海市优秀工程咨询成果三等奖
  Award: Third-class Prize of Shanghai for Outstanding Construction Project Consulting in 2011

- **盛大全球研发中心—多层科研楼建筑节能评估报告**
  Building Energy Performance Evaluation Report of Multistorey Scientific Research Building: SHANDA Global Research and Development Center
  2011上海市优秀工程咨询成果三等奖
  Award: Third-class Prize of Shanghai for Outstanding Construction Project Consulting in 2011

- **路发广场建筑节能评估报告**
  Building Energy Performance Evaluation Report of Lufa Plaza
  2011上海市优秀工程咨询成果三等奖
  Award: Third-class Prize of Shanghai for Outstanding Construction Project Consulting in 2011

- **由由幼儿园新建工程建筑节能评估报告**
  Building Energy Performance Evaluation Report of New Construction of Youyou Kindergarten
  2011上海市优秀工程咨询成果三等奖
  Award: Third-class Prize of Shanghai for Outstanding Construction Project Consulting in 2011

2010

- **上海世博会浦东临时场馆及配套设施景观设计方案编制**
  Conceptual Landscape Design of Temporary Halls and Facilities in Pudong District for Shanghai Expo
  2010上海市优秀工程咨询成果一等奖
  Award: First-class Prize of Shanghai for Outstanding Construction Project Consulting in 2010

- **上海市浦东新区高东福利院工程可行性研究报告**
  Feasibility Study Report of Gaodong Welfare Home Construction Project in Pudong New District of Shanghai
  2010上海市优秀工程咨询成果二等奖
  Award: Second-class Prize of Shanghai for Outstanding Construction Project Consulting in 2010

- **浦东南路（浦电路—上南路）、耀华路（上南路—浦电路）改建工程可研报告**
  Feasibility Study Report of Reconstruction of Pudong South Road (from Pudian Road to Shangnan Road) and Yaohua Road (from Shangnan Road to Pudian Road)
  2010上海市优秀工程咨询成果二等奖
  Award: Second-class Prize of Shanghai for Outstanding Construction Project Consulting in 2010

工程咨询
ENGINEERING CONSULTING

2010

- **曹路镇第二期自然村落污水纳管工程可行性研究报告**
  Feasibility Study Report of the Second Phase Sewage-into-Sewer Construction Project of Existing Villages in Caolu Town
  2010上海市优秀工程咨询成果二等奖
  Award: Second-class Prize of Shanghai for Outstanding Construction Project Consulting in 2010

- **张衡绿地（张衡公园）景观工程可行性研究报告**
  Feasibility Study Report of Landscape Construction Project of Zhangheng Park
  2010上海市优秀工程咨询成果二等奖
  Award: Second-class Prize of Shanghai for Outstanding Construction Project Consulting in 2010

- **闻居路（西乐路—凌空路）新建工程可行性研究报告**
  Feasibility Study Report of New Construction of Wenju Road (from Xile Road to Lingkong Road)
  2010上海市优秀工程咨询成果三等奖
  Award: Third-class Prize of Shanghai for Outstanding Construction Project Consulting in 2010

- **杨高路（环南一大道—龙阳路立交）改造工程可行性研究报告**
  Feasibility Study Report of Yanggao Road Renovation Project (from Huannan No.1 Avenue to Longyang Road Crossway)
  2010上海市优秀工程咨询成果三等奖
  Award: Third-class Prize of Shanghai for Outstanding Construction Project Consulting in 2010

- **长清路（成山路—杨高路）扩建工程可行性研究报告**
  Feasibility Study Report of Changqing Road Extension Project (from Chengshan Road to Yanggao Road)
  2010上海市优秀工程咨询成果三等奖
  Award: Third-class Prize of Shanghai for Outstanding Construction Project Consulting in 2010

- **唐镇新市镇排水系统专业规划**
  Drainage System Planning of Tang Town and Xinshi Town
  2010上海市优秀工程咨询成果三等奖
  Award: Third-class Prize of Shanghai for Outstanding Construction Project Consulting in 2010

- **北蔡集镇雨水泵站新建工程**
  New Construction Project of Rainwater Pump Station in Beicai Town
  2010上海市优秀工程咨询成果三等奖
  Award: Third-class Prize of Shanghai for Outstanding Construction Project Consulting in 2010

- **梅园公园改造工程**
  Renovation Project of Meiyuan Park
  2010上海市优秀工程咨询成果三等奖
  Award: Third-class Prize of Shanghai for Outstanding Construction Project Consulting in 2010

- **曹路镇新农村建设第一期自然村落公建设施改造工程可行性研究报告**
  Feasibility Study Report of Renovation of Public Buildings and Facilities in Original Villages of Caolu Town (the First Phase of New Village Construction Project of Caolu Town)
  2010上海市优秀工程咨询成果三等奖
  Award: Third-class Prize of Shanghai for Outstanding Construction Project Consulting in 2010

工程咨询
ENGINEERING CONSULTING

2010

- **浦东国际机场北通道（申江路—主进场路）道路两侧绿带工程可行性研究报告**
  Feasibility Study Report of Roadside Green Belt Construction Project along the North Road of Pudong International Airport(from Shenjiang Road to the Main Entrance Road)
  2010上海市优秀工程咨询成果三等奖
  Award: Third-class Prize of Shanghai for Outstanding Construction Project Consulting in 2010

- **中环线浦东段（上中路越江隧道—申江路）道路两侧绿带工程可行性研究报告**
  Feasibility Study Report of Roadside Green Belt Construction Project along Middel Ring in Pudong (from Shangzhong Road Crossriver Tunnel to Shenjiang Road)
  2010上海市优秀工程咨询成果三等奖
  Award: Third-class Prize of Shanghai for Outstanding Construction Project Consulting in 2010

- **上海世博会（浦东）协调区建筑与环境综合整治实施导则**
  Implementation Guidelines for the Comprehensive Improvement of Buildings and Environment in Shanghai Expo (Pudong) Coordination Area
  2010上海市优秀工程咨询成果二等奖
  Award: Second-class Prize of Shanghai for Outstanding Construction Project Consulting in 2010

2009

- **陆家嘴功能区域道路交通建设专题研究**
  Subject Research on Road Network and Traffic System Construction in Lujiazui Region
  2009上海市优秀工程咨询成果二等奖
  Award: Second-class Prize of Shanghai for Outstanding Construction Project Consulting in 2009

- **世博会规划区周边路网道路交通规划实施方案**
  Implementation Scheme of Surrounding Road Network and Traffic Planning for the Region of Shanghai Expo
  2009上海市优秀工程咨询成果二等奖
  Award: Second-class Prize of Shanghai for Outstanding Construction Project Consulting in 2009

- **浦东新区川杨河桥梁规划方案**
  Conceptual Planning of Bridges across Chuanyang River in Pudong New District
  2009上海市优秀工程咨询成果三等奖
  Award: Third-class Prize of Shanghai for Outstanding Construction Project Consulting in 2009

上海浦东建筑设计研究院有限公司部分获奖项目
Part of Prize-winning Projects Made by PDAD

项目名称:上海浦东高东福利院
获奖级别:上海市优秀勘察设计三等奖
颁奖单位:上海市勘察设计行业协会
Project Name:Welfare Home of Gaodong in Pudong of Shanghai
Award:Third-class Prize of Shanghai for Outstanding Exploration and Design
Awarded by:Shanghai Exploration and Design Trade Association

项目名称:上海世纪公园七号门餐厅扩建工程
获奖级别:上海市优秀勘察设计三等奖
颁奖单位:上海市勘察设计行业协会
Project Name:Expansion Project of the Restaurant at the Gate No.7 of Shanghai Century Park
Award:Third-class Prize of Shanghai for Outstanding Exploration and Design
Awarded by:Shanghai Exploration and Design Trade Association

项目名称:莎车县综合福利中心
获奖级别:上海市优秀勘察设计三等奖
颁奖单位:上海市勘察设计行业协会
Project Name:Shache Welfare Center
Award:Third-class Prize of Shanghai for Outstanding Exploration and Design
Awarded by:Shanghai Exploration and Design Trade Association

项目名称:上海市中环线浦东段(上中路越江隧道—申江路)设计6标(罗山路立交)工程
获奖级别:上海市优秀勘察设计三等奖
颁奖单位:上海市勘察设计行业协会
Project Name:the Construction of the Middle Ring in Pudong District
(from Shangzhong Road Cross-river Tunnel to Shenjiang Road)
and the Sixth Section Design(Luoshan Road Overpass)
Award:Third-class Prize of Shanghai for Outstanding Exploration and Design
Awarded by:Shanghai Exploration and Design Trade Association

项目名称:新疆莎车县站前路(米夏路-艾斯提皮尔路)新建工程
获奖级别:上海市优秀勘察设计三等奖
颁奖单位:上海市勘察设计行业协会
Project Name:New Construction of Zhanqian Road of Shache in Xinjiang
(from Mixia Road to Estin peel Road)
Award:Third-class Prize of Shanghai for Outstanding Exploration and Design
Awarded by:Shanghai Exploration and Design Trade Association

项目名称:乌鲁木齐市头屯河区工业大道道路新建工程王家沟大桥
获奖级别:上海市优秀勘察设计三等奖
颁奖单位:上海市勘察设计行业协会
Project Name:New Construction of Gongye Avenue of Toutunhe District in Urumqi
(Wangjiagou Bridge)
Award:Third-class Prize of Shanghai for Outstanding Exploration and Design
Awarded by:Shanghai Exploration and Design Trade Association

项目名称:贵阳市西二环道路工程(甲秀中路、甲秀北路)
获奖级别:上海市优秀勘察设计三等奖
颁奖单位:上海市勘察设计行业协会
Project Name:Construction of West Second Ring Road in Guiyang
(Middle Jiaxiu Road and North Jiaxiu Road)
Award:Third-class Prize of Shanghai for Outstanding Exploration and Design
Awarded by:Shanghai Exploration and Design Trade Association

项目名称:上海市杨思东块雨水泵站新建工程
获奖级别:上海市优秀勘察设计三等奖
颁奖单位:上海市勘察设计行业协会
Project Name:New Construction of Rain Pumping Station in the East Block
of Yangsi in Shanghai
Award:Third-class Prize of Shanghai for Outstanding Exploration and Design
Awarded by:Shanghai Exploration and Design Trade Association

项目名称:上海张衡公园景观工程
获奖级别:上海市优秀勘察设计三等奖
颁奖单位:上海市勘察设计行业协会
Project Name:Landscape Project of Zhangheng Park in Shanghai
Award:Third-class Prize of Shanghai for
Outstanding Exploration and Design
Awarded by:Shanghai Exploration and Design Trade Association

项目名称:上海三林老街城市公园
获奖级别:上海市优秀勘察设计三等奖
颁奖单位:上海市勘察设计行业协会
Project Name:Landscape Project of the Laojie Urban Park in Sanlin District
Award:Third-class Prize of
Shanghai for Outstanding Exploration and Design
Awarded by:Shanghai Exploration and Design Trade Association

项目名称:上海中环线浦东段(上中路越江隧道—大寨河桥)道路及两侧绿带景观工程
获奖级别:上海市优秀勘察设计三等奖
颁奖单位:上海市勘察设计行业协会
Project Name:Landscape Project of the Middle Ring in Pudong District
(from Shangzhong Road Cross-river Tunnel to Dazhai River)
Award:Third-class Prize of Shanghai for Outstanding Exploration and Design
Awarded by:Shanghai Exploration and Design Trade Association

项目名称:世博公园A区(亩中山水)
获奖级别:上海市优秀工程设计一等奖
颁奖单位:上海市勘察设计行业协会
Project Name:Area A in Shanghai Expo Park:Mountain and River in a Plot
Award:First-class Prize of Shanghai for Outstanding Construction and Design
Awarded by:Shanghai Exploration and Design Trade Association

项目名称:上海中环线浦东段(上中路越江隧道—申江路)新建工程
获奖级别:上海市优秀工程设计一等奖
颁奖单位:上海市勘察设计行业协会
Project Name:New Construction of the Middle Ring
in Pudong District, Shanghai (from Shangzhong
Road Cross River Tunnel to Shenjiang Road)
Award:First-class Prize of Shanghai for Outstanding Construction and Design
Awarded by:Shanghai Exploration and Design Trade Association

项目名称:梅园公园改造工程
获奖级别:上海市优秀工程设计三等奖
颁奖单位:上海市勘察设计行业协会
Project Name:Renovation of Meiyuan Park
Award:Third-class Prize of Shanghai for Outstanding Construction and Design
Awarded by:Shanghai Exploration and Design Trade Association

项目名称:浦东南路景观综合改造工程
获奖级别:上海市优秀工程设计二等奖
颁奖单位:上海市勘察设计行业协会
Project Name:Landscape Intergratedrenovation of South Pudong Road
Award:Second-class Prize of Shanghai for
Outstanding Construction and Design
Awarded by:Shanghai Exploration and Design Trade Association

项目名称:上海世博会浦东场地公共空间设计
获奖级别:第一届优秀风景园林规划设计二等奖
颁奖单位:中国风景园林学会
Project Name:Public Space Planning of Pudong Zone for Shanghai Expo.
Award:Second-class Prize in the First Landscape Architectural Design
and Planning Contest
Awarded by:Chinese Society of Landscape Architecture

项目名称：上南路建筑与环境综合整治工程
获奖级别：上海市优秀工程设计三等奖
颁奖单位：上海市勘察设计行业协会
Project Name：Architectural and Environmental Renewal of Shangnan Road
Award：Third-class Prize of Shanghai for Outstanding Design
Awarded by：Shanghai Exploration and Design Trade Association

项目名称：罗山路（龙东大道—S20）快速化改建工程
获奖级别：上海市优秀工程咨询成果二等奖
颁奖单位：上海市工程咨询行业协会
Project Name：Reconstruction Project of Luoshan Express Way (Longdong Avenue-S20)
Award：Second-class Prize of Shanghai for Outstanding Construction Project Consulting
Awarded by：Shanghai Construction Consultants Association

项目名称：浦东南路景观综合改造方案
获奖级别：第一届优秀风景园林规划设计三等奖
颁奖单位：中国风景园林学会
Project Name：Landscape Renovation of Pudong South Road
Award：Third-class Prize in the First Landscape Architectural Design
and Planning Contest
Awarded by：Chinese Society of Landscape Architecture

项目名称：低碳·彩虹谷-山西吕梁离石区北川河片区修建性详细规划
获奖级别：年全国人居经典建筑规划设计方案建筑、科技双金奖
颁奖单位：中国建筑学会
Project Name：Low-carbon Rainbow Valley：
Constructive Detailed Planning of Beichuan River Region，
Lishi District, Lvliang, Shanxi Province
Award：National Gold Award for Outstanding Architectural Design
and Technology in Residence
Awarded by：Architectural Society of China

项目名称：上海世博中国园"亩中山水"设计
获奖级别：第一届优秀风景园林规划设计二等奖
颁奖单位：中国风景园林学会
Project Name：Mountain and River in a Plot：Chinese Park Design in Shanghai Expo Region
Award：Second-class Prize in the First Landscape Architectural Design and Planning Contest
Awarded by：Chinese Society of Landscape Architecture

项目名称：曹路镇第二期自然村落污水纳管工程可行性研究报告
获奖级别：上海市优秀工程咨询成果二等奖
颁奖单位：上海市工程咨询行业协会
Project Name：Feasibility Study Report of the Second Phase
Sewage-into-Sewer Construction Project of Existing villages in Caolu Town
Award：Second-class Prize of Shanghai for Outstanding Construction Project Consulting
Awarded by：Shanghai Construction Consultants Association

项目名称：北京石景山万达广场商业步行街室内设计
获奖级别：上海市建筑装饰奖
颁奖单位：上海装饰装修行业协会
Project Name：Interior Design of Commercial Walking Street of Wanda Plaza
in Shijingshan District, Beijing
Award：Architectural Decoration Award of Shanghai
Awarded by：Shanghai Decoration Association

项目名称：上海万达商业广场地下一层购物中心室内装饰工程
获奖级别：上海市建筑装饰奖
颁奖单位：上海装饰装修行业协会
Project Name：Interior Decoration Project of Underground Floor One Shopping
Center in Wanda Plaza, Shanghai
Award：Architectural Decoration Award of Shanghai
Awarded by：Shanghai Decoration Association

项目名称：吉瑞泰盛国际生活广场售楼处
获奖级别：上海市建筑装饰奖
颁奖单位：上海装饰装修行业协会
Project Name：Sales Office of Jerry Tightsen International Living Mall
Award：Architectural Decoration Award of Shanghai
Awarded by：Shanghai Decoration Association

项目名称：上海爱法小天地酒店式公寓装修设计
获奖级别：上海市建筑装饰奖
颁奖单位：上海装饰装修行业协会
Project Name：Decoration Design of "Shanghai Lovely France：
Little World" Service Apartment
Award：Architectural Decoration Award of Shanghai
Awarded by：Shanghai Decoration Association

项目名称：西安万达广场步行街内装设计
获奖级别：上海市建筑装饰奖
颁奖单位：上海装饰装修行业协会
Project Name：Interior Decoration Design of Wanda Plaza Walking Street in Xi'an
Award：Architectural Decoration Award of Shanghai
Awarded by：Shanghai Decoration Association

项目名称：无锡爱家—金河湾
获奖级别：上海市优秀住宅工程小区设计二等奖
颁奖单位：上海市勘察设计行业协会
Project Name：Lovely Home in Wuxi：Golden River Bend
Award：Second-class Prize of
Shanghai for Outstanding Residence Design of Neighborhoods
Awarded by：Shanghai Exploration and Design Trade Association

项目名称：长清路（成山路—杨高路）扩建工程可行性研究报告
获奖级别：上海市优秀工程咨询成果三等奖
颁奖单位：上海市工程咨询行业协会
Project Name：Feasibility Study Report of Changqing Road Extension Project
(from Chengshan Road to Yanggao Road)
Award：Third-class Prize of Shanghai for Outstanding Construction Project Consulting
Awarded by：Shanghai Construction Consultants Association

项目名称：上海世博会浦东临时场馆及配套设施景观设计方案编制
获奖级别：上海市优秀工程咨询成果一等奖
颁奖单位：上海市工程咨询行业协会
Project Name：Conceptual Landscape Design of Temporary Halls
and Facilities in Pudong for Shanghai Expo
Award：First-class Prize of Shanghai for Outstanding Construction Project Consulting
Awarded by：Shanghai Construction Consultants Association

项目名称：杨高路（环南一大道—龙阳路立交）改造工程可行性研究报告
获奖级别：上海市优秀工程咨询成果三等奖
颁奖单位：上海市工程咨询行业协会
Project Name：Feasibility Study Report of Yanggao Road Renovation Project
(from Huannan No.1 Avenue to Longyang Road Crossway)
Award：Third-class Prize of Shanghai for Outstanding Construction Project Consulting
Awarded by：Shanghai Construction Consultants Association

项目名称：浦东南路（浦电路—上南路）、耀华路（上南路—浦电路）改建工程可研报告
获奖级别：上海市优秀工程咨询成果二等奖
颁奖单位：上海市工程咨询行业协会
Project Name：Feasibility Study Report of Reconstruction of Pudong South Road
(from Pudian Road to Shangnan Road)
and Yaohua Road (from Shangnan Road to Pudian Road)
Award：Second-class Prize of Shanghai for Outstanding Construction Project Consulting
Awarded by：Shanghai Construction Consultants Association

项目名称：浦东新区生态专项建设工程（金海湿地公园）详细规划
获奖级别：上海市优秀规划设计三等奖
颁奖单位：上海市规划设计协会
Project Name：Detailed Planning of Jinhai Wetland Park：
Special Ecological Renovation Project of Pudong New District
Award：Third-class Prize of Shanghai for Outstanding Urban Design
Awarded by：Shanghai Urban Planning and Design Association

项目名称：川沙新镇福利院
获奖级别：上海市优秀工程设计三等奖
颁奖单位：上海市勘察设计协会
Project Name：Welfare Home of Chuansha New Town
Award：Third-class Prize of Shanghai for Outstanding Construction and Design
Awarded by：Shanghai Exploration and Design Trade Association

项目名称:五洲大道(浦东北路—外环线)工程
获奖级别:上海市优秀工程设计一等奖
颁奖单位:上海市勘察设计协会
Project Name:Wuzhou Avenue (from Pudong North Road to Outer Ring) Project
Award:First-class Prize of Shanghai for Outstanding Construction and Design
Awarded by:Shanghai Exploration and Design Trade Association

项目名称:外高桥功能区域市政建设项目储备研究
获奖级别:上海市优秀工程咨询成果二等奖
颁奖单位:上海市工程咨询行业协会
Project Name:Study on Reservation of Municipal Construction Projects in Waigaoqiao District
Award:Second-class Prize of Shanghai for Outstanding Construction Project Consulting
Awarded by:Shanghai Construction Consultants Association

项目名称:西宁煌水森林公园主题园区景观工程
获奖级别:上海市优秀工程设计三等奖
颁奖单位:上海市勘察设计协会
Project Name:Landscape Project of Theme Park in Huangshui Forest Park of Xining
Award:Third-class Prize of Shanghai for Outstanding Construction and Design
Awarded by:Shanghai Exploration and Design Trade Association

项目名称:江苏省徐州市行政中心市民中心绿化环境景观工程
获奖级别:上海市优秀工程设计二等奖
颁奖单位:上海市勘察设计协会
Project Name:Virescence Landscape Project of the plaza of Administration and Civic Center in Xuzhou, Jiangsu Province
Award:Second-class Prize of Shanghai for Outstanding Construction and Design
Awarded by:Shanghai Exploration and Design Trade Association

项目名称:上海应用技术学院奉贤新校区市政配套工程
获奖级别:上海市优秀工程设计三等奖
颁奖单位:上海市勘察设计协会
Project Name:Municipal Facility Project of Fengxian New Campus of Shanghai Institute of Technology
Award:Third-class Prize of Shanghai for Outstanding Construction and Design
Awarded by:Shanghai Exploration and Design Trade Association

项目名称:川沙公园改造工程
获奖级别:上海市优秀工程设计三等奖
颁奖单位:上海市勘察设计协会
Project Name: Renovation Project of Chuansha Park
Award:Third-class Prize of Shanghai for Outstanding Construction and Design
Awarded by:Shanghai Exploration and Design Trade Association

项目名称:上海世博会临时展馆与配套设施景观设计方案编制
获奖级别:上海市优秀城乡规划设计二等奖
颁奖单位:上海市规划设计协会
Project Name:Conceptual Landscape Design of Temporary Halls and Facilities for Shanghai Expo
Award:Second-class Prize of Shanghai for Outstanding Urban Design
Awarded by:Shanghai Urban Planning and Design Association

项目名称:上海世博会临时展馆与配套设施景观设计方案编制
获奖级别:全国优秀城乡规划设计三等奖
颁奖单位:中国城市规划协会
Project Name:Conceptual Landscape Design of Temporary Halls and Facilities for Shanghai Expo
Award:National third-class Prize for Outstanding Urban Design
Awarded by:China Association of City Planning

项目名称:五洲大道(浦东北路—外环线)工程
获奖级别:国家优质工程银奖
颁奖单位:国家工程建设质量奖审定委员会
Project Name:Wuzhou Avenue (From Pudong North Road to Outer Ring) Project
Award:National Silver Award for Outstanding Construction
Awarded by:Evaluation Committee for State Outstanding Construction Quality Award

项目名称:五洲大道(浦东北路—外环线)工程
获奖级别:全国优秀工程勘察设计行业奖、市政公用工程三等奖
颁奖单位:中国勘察设计协会
Project Name:Wuzhou Avenue (from Pudong North Road to Outer Ring) Project
Award:National Prize for Outstanding Exploration and Design, National third-class Prize for Municipal Facility Design
Awarded by:Chinese Exploration and Design Association

项目名称:东郊庄园(原通州市水木天成花苑)
获奖级别:上海市优秀住宅工程小区设计三等奖
颁奖单位:上海市勘察设计协会
Project Name:Eastern Suburb Manor: Garden with Naturally Formed Rivers and Trees in Tongzhou
Award:Third-class Prize of Shanghai for Outstanding Housing Project and Residential Design
Awarded by:Shanghai Exploration and Design Trade Association

项目名称:川沙新市镇排水系统专业规划
获奖级别:上海市优秀城乡规划设计三等奖
颁奖单位:上海市规划行业协会
Project Name:Drainage System Planning of Chuansha New Town
Award:Third-class Prize of Shanghai for Outstanding Urban Planning
Awarded by:Shanghai Urban Planning Trade Association

项目名称:锦博苑动迁商品住宅
获奖级别:上海市优秀住宅工程小区设计三等奖
颁奖单位:上海市勘察设计协会
Project Name:Resettlement Commodity Housing of Jinbo Garden
Award:Third-class Prize of Shanghai for Outstanding Housing Project and Residential Design
Awarded by:Shanghai Exploration and Design Trade Association

项目名称:浦东市民中心
获奖级别:上海市优秀勘察设计一等奖
颁奖单位:上海市勘察设计协会
Project Name:Civic Center of Pudong District
Award:First-class Prize of Shanghai for Outstanding Exploration and Design
Awarded by:Shanghai Exploration and Design Trade Association

项目名称:五洲大道(浦东北路—外环线)工程
获奖级别:上海市优秀勘察设计一等奖
颁奖单位:上海市勘察设计协会
Project Name:Wuzhou Avenue Project (from North Pudong Road to Outer Ring)
Award:First-class Prize of Shanghai for Outstanding Exploration and Design
Awarded by:Shanghai Exploration and Design Trade Association

项目名称:诸暨铁路新客站
获奖级别:上海市优秀勘察设计三等奖
颁奖单位:上海市勘察设计协会
Project Name:New Railway Station of Zhuji
Award:Third-class Prize of Shanghai for Outstanding Exploration and Design
Awarded by:Shanghai Exploration and Design Trade Association

项目名称:浙江省德清县北湖街延伸工程(09省道城区段改造工程)
获奖级别:上海市优秀勘察设计三等奖
颁奖单位:上海市勘察设计协会
Project Name:North Lake Avenue Extension Project in Deqing County,Zhejiang Province
Award:Third-class Prize of Shanghai for Outstanding Exploration and Design
Awarded by:Shanghai Exploration and Design Trade Association

项目名称:湖南电力科技园景观工程
获奖级别:上海市优秀勘察设计三等奖
颁奖单位:上海市勘察设计协会
Project Name:Landscape Project of Electrical Science and Technology Park in Hunan Province
Award:Third-class Prize of Shanghai for Outstanding Exploration and Design
Awarded by:Shanghai Exploration and Design Trade Association

项目名称:芳甸路（锦绣路—杨高路）绿化工程
获奖级别:上海市优秀勘察设计专业三等奖
颁奖单位:上海市勘察设计协会
Project Name:Virescence Engineering of
Fangdian Road (from Jinxiu Road to Yanggao Road)
Award:Third-class Prize of Shanghai for
Outstanding Exploration and Design
Awarded by:Shanghai Exploration and Design Trade Association

项目名称:临港新城申港、临港大道景观绿化工程
获奖级别:上海市优秀勘察设计专业一等奖
颁奖单位:上海市勘察设计协会
Project Name:Landscape Design of Shengang Road
and Lingang Road in Lingang New Town
Award:First-class Prize of Shanghai for Outstanding Exploration and Design
Awarded by:Shanghai Exploration and Design Trade Association

项目名称:三林世博家园公共绿地一期建设工程
获奖级别:上海市优秀勘察设计三等奖
颁奖单位:上海市勘察设计协会
Project Name:The First Phase Construction Project of
the Public Green Space in Sanlin Expo Home
Award:Third-class Prize of Shanghai for Outstanding Exploration and Design
Awarded by:Shanghai Exploration and Design Trade Association

项目名称:锦博苑动迁商品住宅工程
获奖级别:上海市优秀住宅小区创优设计项目优良奖
颁奖单位:上海市勘察设计协会
Project Name:Project for Resettlement Commodity Housing of Jinbo Garden
Award:Excellent Prize of Shanghai for Outstanding Residence
Community Planning and Design
Awarded by:Shanghai Exploration and Design Trade Association

项目名称:国家信息安全工程研究中心工程
获奖级别:军队一等奖
颁奖单位:中国人民解放军总后勤部
Project Name：Project for Research Center of National Information Security Engineering
Award:First-class Award of the PLA Army
Awarded by:Chinese PLA General Logistics Department

项目名称:源远流长
获奖级别:上海国际立体花坛大赛上海地区银奖
颁奖单位:上海国际立体花坛大赛组委会
Project Name:of a Long Standing
Award:Silver Award of Shanghai Region for Excellent Flower Bed Design in
Shanghai International Flower Bed Design Competition
Awarded by:Committee of International Flower Bed Design Contest in Shanghai

项目名称:浙江省德清县北湖街延伸段(Ⅲ标段)工程（09省道城区段改造工程）
获奖级别:浙江省建设工程钱江杯奖（优质工程）
颁奖单位:浙江省建筑业行业协会
Project Name:North Lake Avenue Extension Project(Section Ⅲ Design)
in Deqing County, Zhejiang Province
(Renovation of Urban Section Form Provincial Highway 09)
Award:Qianjiang Cup of Zhejiang Province for Outstanding Construction
Awarded by:Architectural Trade Association of Zhejiang Province